Student Solutions Manual and Study Guide

Discrete Mathematics with Applications

FOURTH EDITION

Susanna S. Epp
DePaul University

Prepared by

Susanna S. Epp
DePaul University

Tom Jenkyns
Brock University

BROOKS/COLE
CENGAGE Learning

Australia • Brazil • Japan • Korea • Mexico • Singapore • Spain • United Kingdom • United States

For product information and technology assistance, contact us at
**Cengage Learning Customer & Sales Support,
1-800-354-9706**

For permission to use material from this text or product, submit all requests online at **www.cengage.com/permissions**
Further permissions questions can be emailed to
permissionrequest@cengage.com

ISBN-13: 978-0-495-82613-2
ISBN-10: 0-495-82613-8

Brooks/Cole
20 Channel Center Street
Boston, MA 02210
USA

Cengage Learning is a leading provider of customized learning solutions with office locations around the globe, including Singapore, the United Kingdom, Australia, Mexico, Brazil, and Japan. Locate your local office at: **www.cengage.com/global**

Cengage Learning products are represented in Canada by Nelson Education, Ltd.

To learn more about Brooks/Cole, visit
www.cengage.com/brookscole

Purchase any of our products at your local college store or at our preferred online store
www.cengagebrain.com

Printed in the United States of America
1 2 3 4 5 6 7 15 14 13 12 11

Table of Contents

Note to Students

This *Student Solutions Manual and Study Guide* for the fourth edition of *Discrete Mathematics with Applications* contains complete solutions to every third exercise in the text that is not fully answered in Appendix B. It also contains additional explanation, and review material.

The specific topics developed in the book provide a foundation for virtually every other mathematics or computer science subject you might study in the future. Perhaps more important, however, is the book's recurring focus on the logical principles used in mathematical reasoning and the techniques of mathematical proof.

Why should it matter for you to learn these things? The main reason is that these principles and techniques are the foundation for all kinds of careful analyses, whether of mathematical statements, computer programs, legal documents, or other technical writing. A person who understands and knows how to develop basic mathematical proofs has learned to think in a highly disciplined way, is able to deduce correct consequences from a few basic principles, can build a logically connected chain of statements and appreciates the need for giving a valid reason for each statement in the chain, is able to move flexibly between abstract symbols and concrete objects, and can deal with multiple levels of abstraction. Mastery of these skills opens a host of interesting and rewarding possibilities in a person's life.

In studying the subject matter of this book, you are embarking on an exciting and challenging adventure. I wish you much success!

Acknowledgements

I am enormously indebted to the work of Tom Jenkyns, whose eagle eye, mathematical knowledge, and understanding of language made an invaluable contribution to this volume. I am also most grateful to my husband, Helmut Epp and my daughter, Caroline Epp, who constructed all the diagrams and, especially, to my husband, who provided much support and wise counsel over many years.

Susanna S. Epp

Chapter 1: Speaking Mathematically

The aim of this chapter is to provide some of the basic terminology that is used throughout the book. Section 1.1 introduces special terms that are used to describe aspects of mathematical thinking and contains exercises to help you start getting used to expressing mathematical statements both formally and informally. To be successful in mathematics, it is important to be able comfortably to translate from formal to informal and from informal to formal modes of expression. Sections 1.2 and 1.3 introduce the basic notions of sets, relations, and functions.

Section 1.1

6. *a.* s is negative   *b.* negative; the cube root of s is negative (*Or*: $\sqrt[3]{s}$ is negative)

   c. is negative; $\sqrt[3]{s}$ is negative (*Or*: the cube root of s is negative)

9. *a.* have at most two real solutions   *b.* has at most two real solutions   *c.* has at most two real solutions   *d.* is a quadratic equation; has at most two real solutions   *e.* E has at most two real solutions

Section 1.2

6. T_2 and T_{-3} each have two elements, and T_0 and T_1 each have one element.

   Justification: $T_2 = \{2, 2^2\} = \{2, 4\}$, $T_{-3} = \{-3, (-3)^2\} = \{-3, 9\}$,

   $T_1 = \{1, 1^2\} = \{1, 1\} = \{1\}$, and $T_0 = \{0, 0^2\} = \{0, 0\} = \{0\}$.

9. *c.* No: The only elements in $\{1, 2\}$ are 1 and 2, and $\{2\}$ is not equal to either of these.

   d. Yes: $\{3\}$ is one of the elements listed in $\{1, \{2\}, \{3\}\}$.

   e. Yes: $\{1\}$ is the set whose only element is 1.

   g. Yes: The only element in $\{1\}$ is 1, and 1 is an element in $\{1, 2\}$.

   h. No: The only elements in $\{\{1\}, 2\}$ are $\{1\}$ and 2, and 1 is not equal to either of these.

   j. Yes: The only element in $\{1\}$ is 1, which is is an element in $\{1\}$. So every element in $\{1\}$ is in $\{1\}$.

12. All four sets have nine elements.

    a. $S \times T = \{(2,1), (2,3), (2,5), (4,1), (4,3), (4,5), (6,1), (6,3), (6,5)\}$

    b. $T \times S = \{(1,2), (3,2), (5,2), (1,4), (3,4), (5,4), (1,6), (3,6), (5,6)\}$

    c. $S \times S = \{(2,2), (2,4), (2,6), (4,2), (4,4), (4,6), (6,2), (6,4), (6,6)\}$

    d. $T \times T = \{(1,1), (1,3), (1,5), (3,1), (3,3), (3,5), (5,1), (5,3), (5,5)\}$

Section 1.3

6. *a.* $(2, 4) \in R$ because $4 = 2^2$.

   $(4, 2) \notin R$ because $2 \neq 4^2$.

   $(-3, 9) \in R$ because $9 = (-3)^2$.

   $(9, -3) \notin R$ because $-3 \neq 9^2$.

1

b.

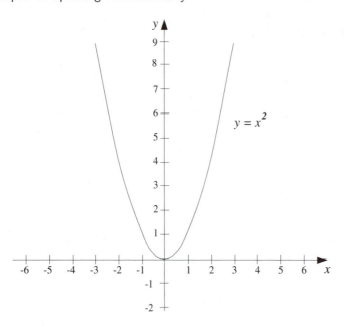

12. T is not a function because, for example, both $(0, 1)$ and $(0, -1)$ are in T but $1 \neq -1$. Many other examples could be given showing that T does not satisfy property (2) of the definition of function.

15. *c.* This diagram does not determine a function because 4 is related to both 1 and 2, which violates property (2) of the definition of function.

   d. This diagram defines a function; both properties (1) and (2) are satisfied.

   e. This diagram does not determine a function because 2 is in the domain but it is not related to any element in the co-domain.

18. $h(-\frac{12}{5}) = h(\frac{0}{1}) = h(\frac{9}{17}) = 2$

Review Guide: Chapter 1

Variables and Mathematical Statements

- What are the two main ways variables are used? *(p. 1)*
- What is a universal statement? Give one example. *(p. 2)*
- What is a conditional statement? Give one example. *(p. 2)*
- What is an existential statement Give one example. *(p. 2)*
- Give an example of a universal conditional statement. *(p. 3)*
- Give an example of a universal existential statement. *(p. 3)*
- Give an example of an existential universal statement. *(p. 4)*

Sets

- What does the notation $x \in S$ mean? *(p. 7)*
- What does the notation $x \notin S$ mean? *(p. 7)*
- How is the set-roster notation used to define a set? *(p. 7)*
- What is the axiom of extension? *(p. 7)*
- What do the symbols **R**, **Z**, and **Q** stand for? *(p. 8)*
- What is the set builder notation? *(p. 8)*
- If S is a set and $P(x)$ is a property that elements may or may not satisfy, how should the following be read out loud: $\{x \in S \mid P(x)\}$? *(p. 8)*

Subsets

- If A and B are sets, what does it mean for A to be a subset of B? What is the notation that indicates that A is a subset of B? *(p. 9)*
- What does the notation $A \nsubseteq B$ mean? *(p. 9)*
- What does it mean for one set to be a proper subset of another? *(p. 9)*
- How are the symbols $\subseteq$ and $\in$ different from each other? *(p. 10)*

Cartesian Products

- What does it mean for an ordered pair (a, b) to equal an ordered pair (c, d)? *(p. 11)*
- Given sets A and B, what is the Cartesian product of A and B.? What is the notation for the Cartesian product of A and B? *(p. 11)*
- What is the Cartesian plane? *(p. 12)*

Relations

- What is a relation from a set A to a set B? *(p. 14)*
- If R is a relation from A to B, what is the domain of R? *(p. 14)*
- If R is a relation from A to B, what is the co-domain of R? *(p. 14)*
- If R is a relation from A to B, what does the notation $x \, R \, y$ mean? *(p. 14)*
- How should the following notation be read: $x \, \not R \, y$? *(p. 14)*
- How is the arrow diagram for a relation drawn? *(p. 16)*

Functions

- What is a function F from a set A to a set B? *(p. 17)*

- What are less formal/more formal ways to state the two properties a function F must satisfy? *(p. 17)*
- Given a function F from a set A to a set B and an element x in A, what is $F(x)$? *(p. 17)*
- What is the squaring function from $\mathbf{R}$ to $\mathbf{R}$? *(p. 20)*
- What is the successor function from $\mathbf{Z}$ to $\mathbf{Z}$? *(p. 20)*
- Give an example of a constant function. *(p. 20)*
- What is the difference between the notations f and $f(x)$? *(pp. 17, 20)*
- If f and g are functions from A to B, what does it mean for f to equal g? *(p. 20)*

Chapter 2: The Logic of Compound Statements

The ability to reason using the principles of logic is essential for solving problems in abstract mathematics and computer science and for understanding the reasoning used in mathematical proof and disproof. In this chapter the various rules used in logical reasoning are developed both symbolically and in the context of their somewhat limited but very important use in everyday language.

Exercise sets for Sections 2.1–2.3 and 3.1–3.4 contain sentences for you to negate, write the contrapositive for, and so forth. These are designed to help you learn to incorporate the rules of logic into your general reasoning processes. Chapters 2 and 3 also present the rudiments of symbolic logic as a foundation for a variety of upper-division courses. Symbolic logic is used in, among others, the study of digital logic circuits, relational databases, artificial intelligence, and program verification.

Section 2.1

9. $(n \lor k) \land \sim (n \land k)$

15. When you are filling out a truth table, a convenient way to remember the definitions of $\sim$ (not), $\land$ (and), and $\lor$ (or) is to think of them in words as follows.

(1) A *not* statement has opposite truth value from that of the statement. So to fill out a column of a truth table for the negation of a statement, you look at the column representing the truth values for the statement. In each row where the truth value for the statement is T, you place an F in the corresponding row in the column representing the truth value of the negation. In each row where the truth value for the statement is F, you place a T in the corresponding row in the column representing the truth value of the negation.

(2) The only time an *and* statement is true is when both components are true. Thus to fill out a column of a truth table where the word *and* links two component statements, just look at the columns with the truth values for the component statements. In any row where both columns have a T, you put a T in the same row in the column for the *and* statement. In every other row of this column, you put a F.

(3) The only time an *or* statement is false is when both components are false. So to fill out a column of a truth table where the word *or* links two component statements, you look at the columns with the truth values for the component statements. In any row where both columns have a F, you put a F in the same row in the column for the *or* statement. In every other row of this column, you put a T.

p	q	r	$\sim q$	$\sim q \lor r$	$p \land (\sim q \lor r)$
T	T	T	F	T	T
T	T	F	F	F	F
T	F	T	T	T	T
T	F	F	T	T	T
F	T	T	F	T	F
F	T	F	F	F	F
F	F	T	T	T	F
F	F	F	T	T	F

The truth table shows that $p \land (q \lor r)$ and $(p \land q) \lor (p \land r)$ always have the same truth values. Therefore they are logically equivalent. This proves the distributive law for $\land$ over $\lor$.

24.

p	q	r	$p \vee q$	$p \wedge r$	$(p \vee q) \vee (p \wedge r)$	$(p \vee q) \wedge r$	
T	T	T	T	T	T	T	
T	T	F	T	F	T	F	$\leftarrow$
T	F	T	T	T	T	T	$\leftarrow$
T	F	F	T	F	T	F	
F	T	T	T	F	T	T	
F	T	F	T	F	T	F	$\leftarrow$
F	F	T	F	F	F	F	
F	F	F	F	F	F	F	

$$\underbrace{\qquad\qquad\qquad\qquad\qquad\qquad\qquad}_{\text{different truth values}}$$

The truth table shows that $(p \vee q) \vee (p \wedge r)$ and $(p \vee q) \wedge r$ have different truth values in rows 2, 3, and 6. Hence they are not logically equivalent.

30. The dollar is not at an all-time high or the stock market is not at a record low.

33. $-10 \geq x$ or $x \geq 2$

39. The statement's logical form is $(p \wedge q) \vee ((r \wedge s) \wedge t)$, so its negation has the form

$$
\begin{aligned}
\sim ((p \wedge q) \vee ((r \wedge s) \wedge t)) \quad &\equiv \quad \sim (p \wedge q) \wedge \sim ((r \wedge s) \wedge t)) \\
&\equiv \quad (\sim p \vee \sim q) \wedge (\sim (r \wedge s) \vee \sim t)) \\
&\equiv \quad (\sim p \vee \sim q) \wedge ((\sim r \vee \sim s) \vee \sim t)).
\end{aligned}
$$

Thus a negation is ($num\_orders \geq 50$ or $num\_instock \leq 300$) and (($50 > num\_orders$ or $num\_orders \geq 75$) or $num\_instock \leq 500$).

42.

p	q	r	$\sim p$	$\sim q$	$\sim p \wedge q$	$q \wedge r$	$((\sim p \wedge q) \wedge (q \wedge r))$	$((\sim p \wedge q) \wedge (q \wedge r)) \wedge \sim q$
T	T	T	F	F	F	T	F	F
T	T	F	F	F	F	F	F	F
T	F	T	F	T	F	F	F	F
T	F	F	F	T	F	F	F	F
F	T	T	T	F	T	T	T	F
F	T	F	T	F	T	F	F	F
F	F	T	T	T	F	F	F	F
F	F	F	T	T	F	F	F	F

$$\underbrace{\qquad\qquad\qquad}_{\text{all } F's}$$

Since all the truth values of $((\sim p \wedge q) \wedge (q \wedge r)) \wedge \sim q$ are F, $((\sim p \wedge q) \wedge (q \wedge r)) \wedge \sim q$ is a contradiction.

45. Let b be "Bob is a double math and computer science major," m be "Ann is a math major," and a be "Ann is a double math and computer science major." Then the two statements can be symbolized as follows: a. $(b \wedge m) \wedge \sim a$ and b. $\sim (b \wedge a) \wedge (m \wedge b)$. *Note*: The entries in the truth table assume that a person who is a double math and computer science major is also a math major and a computer science major.

b	m	a	$\sim a$	$b \wedge m$	$m \wedge b$	$b \wedge a$	$\sim (b \wedge a)$	$(b \wedge m) \wedge \sim a$	$\sim (b \wedge a) \wedge (m \wedge b)$
T	T	T	F	T	T	T	F	F	F
T	T	F	T	T	T	F	T	T	T
T	F	T	F	T	F	T	F	F	F
T	F	F	T	F	F	F	T	F	F
F	T	T	F	F	F	F	T	F	F
F	T	F	T	F	F	F	T	F	F
F	F	T	F	F	F	F	T	F	F
F	F	F	T	F	F	F	T	F	F

same truth values

The truth table shows that $(b \wedge m) \wedge \sim a$ and $\sim (b \wedge a) \wedge (m \wedge b)$ always have the same truth values. Hence they are logically equivalent.

51. *Solution 1:* $\quad p \wedge (\sim q \vee p) \quad \equiv \quad p \wedge (p \vee \sim q) \quad$ commutative law for $\vee$
$$\equiv \quad p \qquad\qquad\qquad\quad \text{absorption law}$$

Solution 2: $\quad p \wedge (\sim q \vee p) \quad \equiv \quad (p \wedge \sim q) \vee (p \wedge p) \quad$ distributive law
$$\equiv \quad (p \wedge \sim q) \vee p \qquad \text{identity law for } \wedge$$
$$\equiv \quad p \qquad\qquad\qquad\quad \text{by exercise 50.}$$

54. $\quad (p \wedge (\sim (\sim p \vee q))) \vee (p \wedge q) \quad \equiv \quad (p \wedge (\sim (\sim p) \wedge \sim q)) \vee (p \wedge q) \qquad$ De Morgan's law
$$\equiv \quad (p \wedge (p \wedge \sim q)) \vee (p \wedge q) \qquad \text{double negative law}$$
$$\equiv \quad ((p \wedge p) \wedge \sim q)) \vee (p \wedge q) \qquad \text{associative law for } \wedge$$
$$\equiv \quad (p \wedge \sim q)) \vee (p \wedge q) \qquad\quad \text{idempotent law for } \wedge$$
$$\equiv \quad p \wedge (\sim q \vee q) \qquad\qquad\quad \text{distributive law}$$
$$\equiv \quad p \wedge (q \vee \sim q) \qquad\qquad\quad \text{commutative law for } \vee$$
$$\equiv \quad p \wedge \mathbf{t} \qquad\qquad\qquad\quad\; \text{negation law for } \vee$$
$$\equiv \quad p \qquad\qquad\qquad\qquad\;\; \text{identity law for } \wedge$$

Section 2.2

6.

p	q	$\sim p$	$\sim p \wedge q$	$p \vee q$	$(p \vee q) \vee (\sim p \wedge q)$	$(p \vee q) \vee (\sim p \wedge q) \to q$
T	T	F	F	T	T	T
T	F	F	F	T	T	F
F	T	T	T	T	T	T
F	F	T	F	F	F	T

15.

p	q	r	$q \to r$	$p \to q$	$p \to (q \to r)$	$(p \to q) \to r$	
T	T	T	T	T	T	T	
T	T	F	F	T	F	F	
T	F	T	T	F	T	T	
T	F	F	T	F	T	T	
F	T	T	T	T	T	T	
F	T	F	F	T	T	F	$\leftarrow$
F	F	T	T	T	T	F	$\leftarrow$
F	F	F	T	T	T	F	$\leftarrow$

different truth values

The truth table shows that $p \to (q \to r)$ and $(p \to q) \to r$ do not always have the same truth values. (They differ for the combinations of truth values for p, q, and r shown in rows 6, 7, and 8.) Therefore they are not logically equivalent.

18. *Part 1*: Let p represent "It walks like a duck," q represent "It talks like a duck," and r represent "It is a duck." The statement "If it walks like a duck and it talks like a duck, then it is a duck" has the form $p \wedge q \rightarrow r$. And the statement "Either it does not walk like a duck or it does not talk like a duck or it is a duck" has the form $\sim p \vee \sim q \vee r$.

p	q	r	$\sim p$	$\sim q$	$p \wedge q$	$\sim p \vee \sim q$	$p \wedge q \rightarrow r$	$(\sim p \vee \sim q) \vee r$
T	T	T	F	F	T	F	T	T
T	T	F	F	F	T	F	F	F
T	F	T	F	T	F	T	T	T
T	F	F	F	T	F	T	T	T
F	T	T	T	F	F	T	T	T
F	T	F	T	F	F	T	T	T
F	F	T	T	T	F	T	T	T
F	F	F	T	T	F	T	T	T

$$\underbrace{\qquad\qquad\qquad\qquad\qquad}_{\text{same truth values}}$$

The truth table shows that $p \wedge q \rightarrow r$ and $(\sim p \vee \sim q) \vee r$ always have the same truth values. Thus the following statements are logically equivalent: "If it walks like a duck and it talks like a duck, then it is a duck" and "Either it does not walk like a duck or it does not talk like a duck or it is a duck."

Part 2: The statement "If it does not walk like a duck and it does not talk like a duck then it is not a duck" has the form $\sim p \wedge \sim q \rightarrow \sim r$.

p	q	r	$\sim p$	$\sim q$	$\sim r$	$p \wedge q$	$\sim p \wedge \sim q$	$p \wedge q \rightarrow r$	$(\sim p \wedge \sim q) \rightarrow \sim r$	
T	T	T	F	F	F	T	F	T	T	
T	T	F	F	F	T	T	F	F	T	$\leftarrow$
T	F	T	F	T	F	F	F	T	T	
T	F	F	F	T	T	F	F	T	T	
F	T	T	T	F	F	F	F	T	T	
F	T	F	T	F	T	F	F	T	T	
F	F	T	T	T	F	F	T	T	F	$\leftarrow$
F	F	F	T	T	T	F	T	T	T	

$$\underbrace{\qquad\qquad\qquad\qquad\qquad}_{\text{different truth values}}$$

The truth table shows that $p \wedge q \rightarrow r$ and $(\sim p \wedge \sim q) \rightarrow \sim r$ do not always have the same truth values. (They differ for the combinations of truth values of p, q, and r shown in rows 2 and 7.) Thus they are not logically equivalent, and so the statement "If it walks like a duck and it talks like a duck, then it is a duck" is not logically equivalent to the statement "If it does not walk like a duck and it does not talk like a duck then it is not a duck." In addition, because of the logical equivalence shown in Part 1, we can also conclude that the following two statements are not logically equivalent: "Either it does not walk like a duck or it does not talk like a duck or it is a duck" and "If it does not walk like a duck and it does not talk like a duck then it is not a duck."

21. By the truth table for $\rightarrow$, $p \rightarrow q$ is false if, and only if, p is true and q is false. Under these circumstances, (b) $p \vee q$ is true and (c) $q \rightarrow p$ is also true.

27.

p	q	$\sim p$	$\sim q$	$q \to p$	$\sim p \to \sim q$
T	T	F	F	T	T
T	F	F	T	T	T
F	T	T	F	F	F
F	F	T	T	T	T

same truth values

The truth table shows that $q \to p$ *and* $\sim p \to \sim q$ always have the same truth values, so they are logically equivalent. Thus the converse and inverse of a conditional statement are logically equivalent to each other.

30. The corresponding tautology is $p \wedge (q \vee r) \leftrightarrow (p \wedge q) \vee (p \wedge r)$

p	q	r	$q \vee r$	$p \wedge q$	$p \wedge r$	$p \wedge (q \vee r)$	$(p \wedge q) \vee (p \wedge r)$	$p \wedge (q \vee r) \leftrightarrow$ $(p \wedge q) \vee (p \wedge r)$
T	T	T	T	T	T	T	T	T
T	T	F	T	T	F	T	T	T
T	F	T	T	F	T	T	T	T
T	F	F	F	F	F	F	F	T
F	T	T	T	F	F	F	F	T
F	T	F	T	F	F	F	F	T
F	F	T	T	F	F	F	F	T
F	F	F	F	F	F	F	F	T

all T's

The truth table shows that $p \wedge (q \vee r) \leftrightarrow (p \wedge q) \vee (p \wedge r)$ is always true. Hence it is a tautology.

33. If this integer is even, then it equals twice some integer, and if this integer equals twice some integer, then it is even.

36. The Personnel Director did not lie. By using the phrase "only if," the Personnel Director set forth conditions that were necessary but not sufficient for being hired: if you did not satisfy those conditions then you would not be hired. The Personnel Director's statement said nothing about what would happen if you did satisfy those conditions.

39. *b.* If a security code is not entered, then the door will not open.

45. If this computer program produces error messages during translation, then it is not correct.

If this computer program is correct, then it does not produce error messages during translation.

48. *a.*
$$\begin{aligned} p \vee \sim q \to r \vee q &\equiv \sim (p \vee \sim q) \vee (r \vee q) \qquad \textit{[an acceptable answer]} \\ &\equiv (\sim p \wedge \sim(\sim q)) \vee (r \vee q) \quad \text{by De Morgan's law} \\ &\qquad \qquad \qquad \qquad \qquad \qquad \textit{[another acceptable answer]} \\ &\equiv (\sim p \wedge q) \vee (r \vee q) \qquad \text{by the double negative law} \\ &\qquad \qquad \qquad \qquad \qquad \qquad \textit{[another acceptable answer]} \end{aligned}$$

b.
$$\begin{aligned} p \vee \sim q \to r \vee q &\equiv (\sim p \wedge q) \vee (r \vee q) \qquad \text{by part (a)} \\ &\equiv \sim (\sim(\sim p \wedge q) \wedge \sim (r \vee q)) \quad \text{by De Morgan's law} \\ &\equiv \sim (\sim (\sim p \wedge q) \wedge (\sim r \wedge \sim q)) \quad \text{by De Morgan's law} \end{aligned}$$

The steps in the answer to part (b) would also be acceptable answers for part (a).

51. Yes. As in exercises 47–50, the following logical equivalences can be used to rewrite any statement form in a logically equivalent way using only $\sim$ and $\wedge$:

$$p \to q \equiv \sim p \vee q \qquad \qquad p \leftrightarrow q \equiv (\sim p \vee q) \wedge (\sim q \vee p)$$
$$p \vee q \equiv \sim (\sim p \wedge \sim q) \qquad \qquad \sim (\sim p) \equiv p$$

The logical equivalence $p \wedge q \equiv \ \sim (\sim p \vee \sim q)$ can then be used to rewrite any statement form in a logically equivalent way using only $\sim$ and $\vee$.

Section 2.3

9.

						premises			conclusion	
p	q	r	$\sim q$	$\sim r$	$p \wedge q$	$p \wedge q \to \sim r$	$p \vee \sim q$	$\sim q \to p$	$\sim r$	
T	T	T	F	F	T	F	T	T		
T	T	F	F	T	T	T	T	T	T ←	critical row
T	F	T	T	F	F	T	T	T	F ←	critical row
T	F	F	T	T	F	T	T	T	T ←	critical row
F	T	T	F	F	F	T	F	T		
F	T	F	F	T	F	T	F	T		
F	F	T	T	F	F	T	T	F		
F	F	F	T	T	F	T	T	F		

Rows 2, 3, and 4 of the truth table are the critical rows in which all the premises are true, but row 3 shows that it is possible for an argument of this form to have true premises and a false conclusion. Hence the argument form is invalid.

12. *b.*

		premises		conclusion	
p	q	$p \to q$	$\sim p$	$\sim q$	
T	T	T	F		
T	F	F	F		
F	T	T	T	F ←	critical row
F	F	T	T	T ←	critical row

Rows 3, and 4 of the truth table represent the situations in which all the premises are true, but row 3 shows that it is possible for an argument of this form to have true premises and a false conclusion. Hence the argument form is invalid.

15.

		premise	conclusion	
p	q	q	$p \vee q$	
T	T	T	T ←	critical row
T	F	F		
F	T	T	T ←	critical row
F	F	F		

The truth table shows that in the two situations (represented by rows 1 and 3) in which the premise is true, the conclusion is also true. Therefore, the the second version of generalization is valid.

21.

			premises			conclusion	
p	q	r	$p \vee q$	$p \to r$	$q \to r$	r	
T	T	T	T	T	T	T ←	critical row
T	T	F	T	F	F		
T	F	T	T	T	T	T ←	critical row
T	F	F	T	F	T		
F	T	T	T	T	T	T ←	critical row
F	T	F	T	T	F		
F	F	T	F	T	T		
F	F	F	F	T	T		

The truth table shows that in the three situations (represented by rows 1, 3, 5) in which all three premises are true, the conclusion is also true. Therefore, proof by division into cases is valid.

30. form: $p \to q$ invalid, converse error
          q
      $\therefore$ p

33. A valid argument with a false conclusion must have at least one false premise. In the following example, the second premise is false. (The first premise is true because its hypothesis is false.)

If the square of every real number is positive, then no real number is negative.

The square of every real number is positive.

Therefore, no real number is negative.

42. (1) $q \to r$ premise b
          $\sim r$ premise d
      $\therefore$ $\sim q$ by modus tollens

     (2) $p \vee q$ premise a
          $\sim q$ by (1)
      $\therefore$ p by elimination

     (3) $\sim q \to u \wedge s$ premise e
          $\sim q$ by (1)
      $\therefore$ $u \wedge s$ by modus ponens

     (4) $u \wedge s$ by (3)
      $\therefore$ s by specialization

     (5) p by (2)
          s by (4)
      $\therefore$ $p \wedge s$ by conjunction

     (6) $p \wedge s \to t$ premise c
          $p \wedge s$ by (5)
      $\therefore$ t by modus ponens

Section 2.4

6. The input/output table is as follows:

Input		Output
P	Q	R
1	1	0
1	0	1
0	1	0
0	0	0

12. $(P \vee Q) \vee \sim (Q \wedge R)$

15.

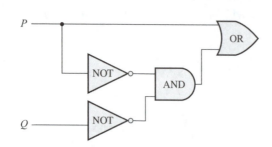

27. The Boolean expression for circuit (a) is $\sim P \wedge (\sim (\sim P \wedge Q))$ and for circuit (b) it is $\sim (P \vee Q)$. We must show that if these expressions are regarded as statement forms, then they are logically equivalent. But

$$
\begin{aligned}
\sim P \wedge (\sim (\sim P \wedge Q)) \quad &\equiv \quad \sim P \wedge (\sim (\sim P) \vee \sim Q) && \text{by De Morgan's law} \\
&\equiv \quad \sim P \wedge (P \vee \sim Q) && \text{by the double negative law} \\
&\equiv \quad (\sim P \wedge P) \vee (\sim P \wedge \sim Q) && \text{by the distributive law} \\
&\equiv \quad \mathbf{c} \vee (\sim P \wedge \sim Q) && \text{by the negation law for } \wedge \\
&\equiv \quad \sim P \wedge \sim Q && \text{by the identity law for } \vee \\
&\equiv \quad \sim (P \vee Q) && \text{by De Morgan's law.}
\end{aligned}
$$

33. *a.*

$$
\begin{aligned}
(P \mid Q) \mid (P \mid Q) \quad &\equiv \quad \sim [(P \mid Q) \wedge (P \mid Q)] && \text{by definition of } \mid \\
&\equiv \quad \sim (P \mid Q) && \text{by the idempotent law for } \wedge \\
&\equiv \quad \sim [\sim (P \wedge Q)] && \text{by definition of } \mid \\
&\equiv \quad P \wedge Q && \text{by the double negative law.}
\end{aligned}
$$

b.

$$
\begin{aligned}
P \wedge (\sim Q \vee R) \quad &\equiv \quad (P \mid (\sim Q \vee R)) \mid (P \mid (\sim Q \vee R)) \\
&\qquad\qquad\qquad\qquad\qquad \text{by part (a)} \\
&\equiv \quad (P \mid [(\sim Q \mid \sim Q) \mid (R \mid R)]) \mid (P \mid [(\sim Q \mid \sim Q) \mid (R \mid R)]) \\
&\qquad\qquad\qquad\qquad\qquad \text{by Example 2.4.7(b)} \\
&\equiv \quad (P \mid [((Q \mid Q) \mid (Q \mid Q)) \mid (R \mid R)]) \mid (P \mid [((Q \mid Q) \mid (Q \mid Q)) \mid (R \mid R)]) \\
&\qquad\qquad\qquad\qquad\qquad \text{by Example 2.4.7(a)}
\end{aligned}
$$

Section 2.5

3. $287 = 256 + 16 + 8 + 4 + 2 + 1 = 100011111_2$

6. $1424 = 1024 + 256 + 128 + 16 = 10110010000_2$

9. $110110_2 = 32 + 16 + 4 + 2 = 54_{10}$

18. $\begin{array}{r} 11010_2 \\ - \quad 1101_2 \\ \hline 1101_2 \end{array}$

21. *b.* $S = 0, T = 1$ *c.* $S = 0, T = 0$

24. $67_{10} = (64 + 2 + 1)_{10} = 01000011_2 \longrightarrow 10111100 \longrightarrow 10111101.$

     So the two's complement is 10111101.

30. $10111010 \longrightarrow -(01000101 + 1)_2 \longrightarrow -01000110_2 = -(64 + 4 + 2)_{10} = -70_{10}$

36. $123_{10} = (64 + 32 + 16 + 8 + 2 + 1)_{10} = 01111011_2$

   $-94_{10} = -(64 + 16 + 8 + 4 + 2)_{10} = -01011110_2 \longrightarrow (10100001 + 1)_2 \longrightarrow 10100010$

   So the 8-bit representations of 123 and -94 are 01111011 and 10100010. Adding the 8-bit representations gives

$$
\begin{array}{r}
01111011 \\
+\quad 10100010 \\
\hline
100011101
\end{array}
$$

   Truncating the 1 in the 2^8th position gives 00011101. Since the leading bit of this number is a 0, the answer is positive. Converting back to decimal form gives

$$00011101 \longrightarrow 11101_2 = (16 + 8 + 4 + 1)_{10} = 29_{10}.$$

   So the answer is 29.

39. $E0D_{16} = 14 \cdot 16^2 + 0 + 13 = 3597_{10}$

42. $B53DF8_{16} = 1011\,0101\,0011\,1101\,1111\,1000_2$

45. $1011\,0111\,1100\,0101_2 = B7C5_{16}$

Review Guide: Chapter 2

Compound Statements

- What is a statement? *(p. 24)*
- If p and q are statements, how do you symbolize "p but q" and "neither p nor q"? *(p. 25)*
- What does the notation $a \leq x < b$ mean? *(p. 26)*
- What is the conjunction of statements p and q? *(p. 27)*
- What is the disjunction of statements p and q? *(p. 28)*
- What are the truth table definitions for $\sim p$, $p \wedge q$, $p \vee q$, $p \to q$, and $p \leftrightarrow q$? *(pp. 26-28,39,45)*
- How do you construct a truth table for a general compound statement form? *(p. 29)*
- What is exclusive or? *(p. 29)*
- What is a tautology, and what is a contradiction? *(p. 34)*
- What is a conditional statement? *(p. 40)*
- Given a conditional statement, what is its hypothesis (antecedent)? conclusion (consequent)? *(p. 39-40)*
- What is a biconditional statement? *(p. 45)*
- What is the order of operations for the logical operators? *(p. 46)*

Logical Equivalence

- What does it mean for two statement forms to be logically equivalent? *(p. 30)*
- How do you test to see whether two statement forms are logically equivalent? *(p. 30)*
- How do you annotate a truth table to explain how it shows that two statement forms are or are not logically equivalent? *(p. 30-31)*
- What is the double negative property? *(p. 31)*
- What are De Morgan's laws? *(p. 32)*
- How is Theorem 2.1.1 used to show that two statement forms are logically equivalent? *(p. 36)*
- What are negations for the following forms of statements? *(pp. 32,42)*
    - $p \wedge q$
    - $p \vee q$
    - $p \to q$ (if p then q)

Converse, Inverse, Contrapositive

- What is the contrapositive of a statement of the form "If p then q"? *(p. 43)*
- What are the converse and inverse of a statement of the form "If p then q"? *(p. 44)*
- Can you express converses, inverses, and contrapositives of conditional statements in ordinary English? *(pp. 43-44)*
- If a conditional statement is true, can its converse also be true? *(p. 44)*
- Given a conditional statement and its contrapositive, converse, and inverse, which of these are logically equivalent and which are not? *(p. 44)*

Necessary and Sufficient Conditions, Only If

- What does it mean to say that something is true only if something else is true? *(p. 45)*
- How are statements about only-if statements translated into if-then form.? *(p. 45)*
- What does it mean to say that something is a necessary condition for something else? *(p. 46)*
- What does it mean to say that something is a sufficient condition for something else? *(p. 46)*

- How are statements about necessary and sufficient conditions translated into if-then form.? *(p. 47)*

Validity and Invalidity

- How do you identify the logical form of an argument? *(p. 24)*
- What does it mean for a form of argument to be valid? *(p. 51)*
- How do you test to see whether a given form of argument is valid? *(p. 52)*
- How do you annotate a truth table to explain how it shows that an argument is or is not valid? *(pp. 52, 59, A9-A10)*
- What does it mean for an argument to be sound? *(p. 59)*
- What are modus ponens and modus tollens? *(pp. 52-53)*
- Can you give examples for and prove the validity of the following forms of argument? *(pp. 54-56)*

$$
\begin{array}{ll}
\dfrac{p}{\therefore \ p \vee q} & \text{and} \qquad \dfrac{q}{\therefore \ p \vee q}
\end{array}
$$

$$
\begin{array}{ll}
\dfrac{p \wedge q}{\therefore \ p} & \text{and} \qquad \dfrac{p \wedge q}{\therefore \ q}
\end{array}
$$

$$
\begin{array}{ll}
\begin{array}{l} p \vee q \\ \sim q \\ \hline \therefore \ p \end{array} & \text{and} \qquad \begin{array}{l} p \vee q \\ \sim p \\ \hline \therefore \ q \end{array}
\end{array}
$$

$$
\begin{array}{l} p \to q \\ q \to r \\ \hline \therefore \ p \to r \end{array}
$$

$$
\begin{array}{l} p \vee q \\ p \to r \\ q \to r \\ \hline \therefore \ r \end{array}
$$

- What are converse error and inverse error? *(pp. 57-58)*
- Can a valid argument have a false conclusion? *(p. 58)*
- Can an invalid argument have a true conclusion? *(p. 59)*
- Which of modus ponens, modus tollens, converse error, and inverse error are valid and which are invalid? *(pp. 53,58)*
- What is the contradiction rule? *(p. 59)*
- How do you use valid forms of argument to solve puzzles such as those of Raymond Smullyan about knights and knaves? *(p. 60)*

Digital Logic Circuits and Boolean Expressions

- Given a digital logic circuit, how do you
    - find the output for a given set of input signals *(p. 68)*
    - construct an input/output table *(pp. 68-69)*
    - find the corresponding Boolean expression? *(pp. 69-70)*
- What is a recognizer? *(p. 70)*
- Given a Boolean expression, how do you draw the corresponding digital logic circuit? *(pp. 70-71)*
- Given an input/output table, how do you draw the corresponding digital logic circuit? *(p. 72)*
- What is disjunctive normal form? *(p. 72)*
- What does it mean for two circuits to be equivalent? *(p. 74)*

- What are NAND and NOR gates? *(p. 74)*
- What are Sheffer strokes and Peirce arrows? *(p. 74)*

Binary and Hexadecimal Notation

- How do you transform positive integers from decimal to binary notation and the reverse? *(pp. 79-80)*
- How do you add and subtract integers using binary notation? *(p. 81)*
- What is a half-adder? *(p. 82)*
- What is a full-adder? *(p. 83)*
- What is the 8-bit two's complement of an integer in binary notation? *(p. 84)*
- How do you find the 8-bit two's complement of a positive integer a that is at most 255? *(p. 85)*
- How do you find the decimal representation of the integer with a given 8-bit two's complement? *(p. 86)*
- How are negative integers represented using two's complements? *(p. 87)*
- How is computer addition with negative integers performed? *(pp. 87-90)*
- How do you transform positive integers from hexadecimal to decimal notation? *(p. 92)*
- How do you transform positive integers from binary to hexadecimal notation and the reverse? *(p. 93)*

- What is octal notation? *(p. 95)*

Chapter 3: The Logic of Quantified Statements

Ability to use the logic of quantified statements correctly is necessary for doing mathematics because mathematics is, in a very broad sense, about quantity. The main purpose of this chapter is to familiarize you with the language of universal and existential statements. The various facts about quantified statements developed in this chapter are used extensively in Chapter 4 and are referred to throughout the rest of the book. Experience with the formalism of quantification is especially useful for students planning to study LISP or Prolog, program verification, or relational databases.

One thing to keep in mind is the tolerance for potential ambiguity in ordinary language, which is typically resolved through context or inflection. For instance, as the "Caution" on page 111 of the text indicates, the sentence "All mathematicians do not wear glasses" is one way to phrase a negation to "All mathematicians wear glasses." (To see this, say it out loud, stressing the word "not.") Some grammarians ask us to avoid such phrasing because of its potentially ambiguity, but the usage is widespread even in formal writing in high-level publications ("All juvenile offenders are not alike," Anthony Lewis, The New York Times, 19 May 1997, Op-Ed page), or in literary works ("All that glisters is not gold," William Shakespeare, The Merchant of Venice, Act 2, Scene 7, 1596-1597).

Even rather complex sentences can be negated in this way. For instance, when asked to write a negation for "The sum of any two irrational numbers is irrational," many people instinctively write "The sum of any two irrational numbers is not irrational." This is an acceptable informal negation (again, say it out loud, stressing the word "not"), but it can lead to genuine mistakes in formal situations. To avoid such mistakes, tell yourself that simply inserting the word "not" is very rarely a good way to express the negation of a mathematical statement.

Section 3.1

6. *a.* When $m = 25$ and $n = 10$,the statement "m is a factor of n^2" is true because $n^2 = 100$ and $100 = 4 \cdot 25$. But the statement "m is a factor of n" is false because 10 is not a product of 25 times any integer. Thus the hypothesis is true and the conclusion is false, so the statement as a whole is false.

b. $R(m, n)$ is also false when $m = 8$ and $n = 4$ because 8 is a factor of $4^2 = 16$, but 8 is not a factor of 4.

c. When $m = 5$ and $n = 10$, both statements "m is a factor of n^2" and "m is a factor of n" are true because $n = 10 = 5 \cdot 2 = m \cdot 2$ and $n^2 = 100 = 5 \cdot 20 = m \cdot 20$. Thus both the hypothesis and conclusion of $R(m, n)$ are true, and so the statement as a whole is true.

d. Here are examples of two kinds of correct answers:

(1) Let $m = 2$ and $n = 6$. Then both statements "m is a factor of n^2" and "m is a factor of n" are true because $n = 6 = 2 \cdot 3 = m \cdot 3$ and $n^2 = 36 = 2 \cdot 18 = m \cdot 18$. Thus both the hypothesis and conclusion of $R(m, n)$ are true, and so the statement as a whole is true.

(2) Let $m = 6$ and $n = 2$. Then both statements "m is a factor of n^2" and "m is a factor of n" are false because $n = 2 \neq 6 \cdot k$, for any integer k, and $n^2 = 4 \neq 6 \cdot j$, for any integer j. Thus both the hypothesis and conclusion of $R(m, n)$ are false, and so the statement as a whole is true.

12. <u>Counterexample:</u> Let $x = 1$ and $y = 1$, and note that

$$\sqrt{x + y} = \sqrt{1 + 1} = \sqrt{2}$$

whereas

$$\sqrt{x} + \sqrt{y} = \sqrt{1} + \sqrt{1} = 1 + 1 = 2,$$

and

$$2 \neq \sqrt{2}.$$

(This is one counterexample among many. Any real numbers x and y with $xy \neq 0$ will produce a counterexample.)

15. *a. Some acceptable answers:* All rectangles are quadrilaterals. If a figure is a rectangle then that figure is a quadrilateral. Every rectangle is a quadrilateral. All figures that are rectangles are quadrilaterals. Any figure that is a rectangle is a quadrilateral.

    b. Some acceptable answers: There is a set with sixteen subsets. Some set has sixteen subsets. Some sets have sixteen subsets. There is at least one set that has sixteen subsets.

18. *c.* $\forall s$, if $C(s)$ then $\sim E(s)$.

    d. $\exists x$ such that $C(s) \wedge M(s)$.

21. *b.* The base angles of T are equal, for any isosceles triangle T.

    d. f is not differentiable, for some continuous function f.

24. *b.* $\exists$ a question x such that x is easy.

    $\exists x$ such that x is a question and x is easy.

27. *c.* This statement translates as "There is a square that is above d." This is false because the only objects above d are a (a triangle) and b (a circle).

    d. This statement translates as "There is a triangle that has f above it," or, "f is above some triangle." This is true because g is a triangle and f is above g.

30. *a.* This statement translates as "There is a prime number that is not odd." This is true. The number 2 is prime and it is not odd.

    c. This statement translates as "There is a number that is both an odd number and a perfect square." This is true. For example, the number 9 is odd and it is also a perfect square (because $9 = 3^2$).

33. *c.* This statement translates as "For all real numbers a and b, if $ab = 0$ then $a = 0$ or $b = 0$," which is true.

    d. This statement translates as "For all real numbers a, b, c, and d, if $a < b$ and $c < d$ then $ac < bd$," which is false.

    <u>Counterexample:</u> Let $a = -2$, $b = 1$, $c = -3$, and $d = 0$.

    Then $a < b$ because $-2 < 1$ and $c < d$ because $-3 < 0$, but $ac \not< bd$ because $ac = (-2)(-3) = 6$ and $bd = 1 \cdot 0 = 0$ and $6 \not< 0$.

Section 3.2

3. *b.* $\exists$ a computer C such that C does not have a CPU.

   d. $\forall$ bands b, b has won fewer than 10 Grammy awards.

6. *b. Formal negation:* $\exists$ a road r on the map such that r connects towns P and Q.

   Some acceptable informal negations: There is a road on the map that connects towns P and Q. Some road on the map connects towns P and Q. Towns P and Q are connected by a road on the map.

12. The proposed negation is not correct. *Correct negation:* There are an irrational number x and a rational number y such that xy is rational. Or: There are an irrational number and a rational number whose product is rational.

15. *b.* True   *d.* True

  e. False: $x = 36$ is a counterexample because the ones digit of x is 6 and the tens digit is neither 1 nor 2.

21. $\exists$ an integer n such that n is divisible by 6 and n is not divisible by 2 and n is not divisible by 3.

24. *b.* If an integer greater than 5 ends in 1, 3, 7, or 9, then the integer is prime.

  If an integer greater than 5 is prime, then the integer ends in 1, 3, 7, or 9.

27. *Converse*: $\forall$ integers d, if $d = 3$ then $6/d$ is an integer.

  Inverse: $\forall$ integers d, if $6/d$ is not an integer, then $d \neq 3$.

  Contrapositive: $\forall$ integers d, if $d \neq 3$ then $6/d$ is not an integer.

  Theconverse and inverse of the statement are both true, but both the statement and its contrapositive are false. For example, when $d = 2$, then $d \neq 3$ but $6/d = 3$ is an integer.

33. *Converse*: If a function is continuous, then it is differentiable.

  Inverse: If a function is not differentiable, then it is not continuous.

  Contrapositive: If a function is not continuous, then it is not differentiable.

  The statement and its contrapositive are true, but both the converse and inverse are false. For example, take the function f defined by $f(x) = |x|$ for all real numbers x. This function is continuous for all real numbers, but it is not differentiable at $x = 0$.

36. *b. One possible answer*: Let $P(x)$ be "$x^2 \neq 2$." The statements "$\forall x \in \mathbf{Z}, x^2 \neq 2$" and "$\forall x \in \mathbf{Q}, x^2 \neq 2$" are true, but the statement "$\forall x \in \mathbf{R}, x^2 \neq 2$" is false.

42. If a person does not pass a comprehensive exam, then that person cannot obtain a master's degree. *Or*: If a person obtains a master's degree then that person passed a comprehensive exam.

Section 3.3

3. *c.* Let $y = \frac{4}{3}$. Then $xy = \left(\frac{3}{4}\right)\left(\frac{4}{3}\right) = 1$.

6. True.

Given $x =$	Choose $y =$	Is y a circle above x, with a different color from x?
e	a, b, or c	yes ✓
g	a or c	yes ✓
h	a or c	yes ✓
j	b	yes ✓

9. *b.* True. *Solution 1*: Let $x = 0$. Then for any real number r, $x + r = r + x = r$ because 0 is an identity for addition of real numbers. Thus, because every element in E is a real number, $\forall y \in E, x + y = y$.

  Solution 2: Let $x = 0$. Then $x + y = y$ is true for each individual element y of E:

Choose $x = 0$	Given $y =$	Is $x + y = y$?
	-2	yes: $0 + (-2) = -2$ ✓
	-1	yes: $0 + (-1) = -1$ ✓
	0	yes: $0 + 0 = 0$ ✓
	1	yes: $0 + 1 = 1$ ✓
	2	yes: $0 + 2 = 2$ ✓

12. *c. first version of negation*: $\exists\, x$ in D such that $\sim (\exists\, y$ in E such that $xy \geq y)$.

final version of negation: $\exists\, x$ in D such that $\forall\, y$ in E, $xy \not\geq y$. (*Or*: $\exists\, x$ in D such that $\forall\, y$ in E, $xy < y$.)

The statement is true. For each number x in D, you can find a y in D so that $xy \geq y$. Here is a table showing one way to do this: how all possible choices for x could be matched with a y so that $xy \geq y$.:

Given $x =$	you could take $y =$	Is $xy \geq y$?
-2	-2	$(-2)\cdot(-2) = 4 \geq -2$ ✓
-1	0	$(-1)\cdot 0 = 0 \geq -1$ ✓
0	1	$0\cdot 1 = 0 \geq 0$ ✓
1	1	$1\cdot 1 = 1 \geq 1$ ✓
2	2	$2\cdot 2 = 4 \geq 2$ ✓

d. first version of negation: $\forall\, x$ in D, $\sim (\forall\, y$ in E, $x \leq y)$.

final version of negation: $\forall x$ in D, $\exists\, y$ in E such that $x \not\leq y$. (*Or*: $\forall x$ in D, $\exists\, y$ in E such that $x > y$.)

The statement is true. It says that there is a number in D that is less than or equal to every number in D. In fact, -2 is in D and -2 is less than or equal to every number in D (-2, -1, 0, 1, and 2).

21. *c.* Statement (1) is true because $x^2 - 2xy + y^2 = (x - y)^2$. Thus given any real number x, take $y = x$, then $x - y = 0$, and so $x^2 - 2xy + y^2 = 0$.

Statement (2) is false. Given any real number x, choose a real number y with $y \neq x$. Then $x^2 - 2xy + y^2 = (x - y)^2 \neq 0$.

d. Statement (1) is true because no matter what real number x might be chosen, y can be taken to be 1 so that $(x - 5)(y - 1) = (x - 5)\cdot 0 = 0$.

Statement (2) is also true. Take $x = 5$. Then for all real numbers y, $(x-5)(y-1) = 0(y-1) = 0$.

e. Statements (1) and (2) are both false because all real numbers have nonnegative squares and the sum of any two nonnegative real numbers is nonnegative. Hence for all real numbers x and y, $x^2 + y^2 \neq -1$.

24. *b.* $\begin{aligned} \sim (\exists x \in D\ (\exists y \in E\ (P(x,y)))) &= \forall x \in D\ (\sim (\exists y \in E\ (P(x,y)))) \\ &= \forall x \in D\ (\forall y \in E\ (\sim P(x,y))) \end{aligned}$

30. *a.* $\forall x \in \mathbf{R}$, $\exists y \in \mathbf{R}^-$ such that $x > y$.

b. The original statement says that there is a real number that is greater than every negative real number. This is true. For instance, 0 is greater than every negative real number.

The statement with interchanged quantifiers says that no matter what real number might be given, it is possible to find a negative real number that is smaller. This is also true. If the number x that is given is positive, y could be taken to be -1. Then $x > y$. On the other hand, if the number x that is given is 0 or negative, y could be taken to be $x - 1$. In this case also, $x > y$.

36. *a.* $\exists$ a person x such that $\forall$ people y, x trusts y.

b. Negation: $\forall$ people x, $\exists$ a person y such that x does not trust y.

Or: Nobody trusts everybody.

45. $\exists! x \in D$ such that $P(x)$ $\equiv$ $\exists x \in D$ such that $(P(x) \wedge (\forall y \in D$, if $P(y)$ then $y = x))$

Or: There exists a unique x in D such that $P(x)$.

Or: There is one and only one x in D such that $P(x)$.

54. *a.* The statement is false. It says that there are a circle and a triangle that have the same color, which is false because all the triangles are blue, and no circles are blue.

   b. $\exists x(\text{Circle}(x) \wedge (\exists y \ (\text{Triangle}(y) \wedge \text{SameColor}(x, y))))$

   c. $\forall x(\sim\text{Circle}(x) \vee \sim (\exists y \ (\text{Triangle}(y) \wedge \text{SameColor}(x, y))))$

   $\equiv \forall x(\sim\text{Circle}(x) \vee (\forall y \ (\sim\text{Triangle}(y) \vee \sim \text{SameColor}(x, y))))$

57. These statements do not necessarily have the same truth values. For example, let $D = \mathbf{Z}$, the set of all integers, let $P(x)$ be "x is even," and let $Q(x)$ be "x is odd." Then the statement "$\forall x \in D, (P(x) \vee Q(x))$" can be written "$\forall$ integers x, x is even or x is odd," which is true. On the other hand, "$(\forall x \in D, P(x)) \vee (\forall x \in D, \ Q(x))$" can be written "All integers are even or all integers are odd," which is false.

60. *a.* No   *b.* No   *c.* $X = g$

Section 3.4

6. This computer program is not correct.

12. invalid, inverse error

15. invalid, converse error

18. valid, universal modus tollens

24. Valid. The only drawing representing the truth of the premises also represents the truth of the conclusion.

27. Valid. The only drawing representing the truth of the premises also represents the truth of the conclusion.

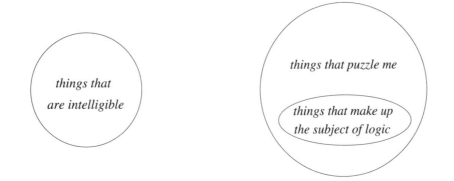

30. 3. If an object is black, then it is a square.

    2. (*contrapositive form*) If an object is a square, then it is above all the gray objects.

    4. If an object is above all the gray objects, then it is above all the triangles.

    1. If an object is above all the triangles, then it is above all the blue objects.

    ∴ If an object is black, then it is above all the blue objects.

36. The universal form of elimination (part a) says that the following form of argument is valid:

    $$\forall x \text{ in } D, \ P(x) \lor Q(x). \qquad \leftarrow major\ premise$$
    $$\sim Q(c) \text{ for a particular } c \text{ in } D. \quad \leftarrow minor\ premise$$
    $$\therefore \quad P(c)$$

    <u>Proof of Validity:</u>

    Suppose the major and minor premises of the above argument form are both true.

    [We must show that the conclusion $P(c)$ is also true.]

    By definition of truth value for a universal statement, $\forall x$ in D, $P(x) \lor Q(x)$ is true if, and only if, the statement "$P(x) \lor Q(x)$" is true for each individual element of D.

    So, by universal instantiation, it is true for the particular element c.

    Hence "$P(c) \lor Q(c)$" is true.

    And since the minor premise says that $\sim Q(c)$, it follows by the elimination rule that $P(c)$ is true.

    [This is what was to be shown.]

Review Guide: Chapter 3

Quantified Statements

- What is a predicate? *(p. 97)*
- What is the truth set of a predicate? *(p. 97)*
- What is a universal statement, and what is required for such a statement to be true? *(p. 98)*
- What is required for a universal statement to be false? *(p. 98)*
- What is the method of exhaustion? *(p. 99)*
- What is an existential statement, and what is required for such a statement to be true? *(p. 99)*
- What is required for a existential statement to be false? *(p. 99)*
- What are some ways to translate quantified statements from formal to informal language? *(p. 100)*
- What are some ways to translate quantified statements from informal to formal language? *(p. 101)*
- What is a universal conditional statement? *(p. 101)*
- What are equivalent ways to write a universal conditional statements? *(pp. 101-103)*
- What are equivalent ways to write existential statements? *(p. 103)*
- What is a trailing quantifier? *(p. 101)*
- What does it mean for a statement to be quantified implicitly? *(p. 103)*
- What do the notations $\Rightarrow$ and $\Leftrightarrow$ mean? *(p. 104)*
- What is the relation among $\forall$, $\exists$, $\wedge$, and $\vee$? *(p. 112)*
- What does it mean for a universal statement to be vacuously true? *(p. 112)*
- What is the rule for interpreting a statement that contains both a universal and an existential quantifier? *(pp. 118-119)*
- How are statements expressed in the computer programming language Prolog? *(pp. 127-128)*

Negations: What are negations for the following forms of statements?

- $\forall x, Q(x)$ *(p. 109)*
- $\exists x$ such that $Q(x)$ *(p. 109)*
- $\forall x$, if $P(x)$ then $Q(x)$ *(p. 111)*
- $\forall x, \exists y$ such that $P(x, y)$ *(p. 123)*
- $\exists x$ such that $\forall y, P(x, y)$ *(p. 123)*

Variants of Conditional Statements

- What are the converse, inverse, and contrapositive of a statement of the form "$\forall x$, if $P(x)$ then $Q(x)$"? *(p. 113)*
- How are quantified statements involving necessary and sufficient conditions and the phrase only-if translated into if-then form? *(pp. 114-115)*

Validity and Invalidity

- What is universal instantiation? *(p. 132)*
- What are the universal versions of modus ponens, modus tollens, converse error, and inverse error, and which of these forms of argument are valid and which are invalid? *(pp. 133-135, 138-139)*
- How is universal modus ponens used in a proof? *(p. 134)*
- How can diagrams be used to test the validity of an argument with quantified statements? *(pp. 136-139)*

Chapter 4: Elementary Number Theory and Methods of Proof

One aim of this chapter is to introduce you to methods for evaluating whether a given mathematical statement is true or false. Throughout the chapter the emphasis is on learning to prove and disprove statements of the form "$\forall x$ in D, if $P(x)$ then $Q(x)$." To prove such a statement directly, you suppose you have a particular but arbitrarily chosen element x in D for which $P(x)$ is true and you show that $Q(x)$ must also be true. To disprove such a statement, you show that there is an element x in D (a counterexample) for which $P(x)$ is true and $Q(x)$ is false. To prove such a statement by contradiction, you show that no counterexample exists, that is, you suppose that there is an x in D for which $P(x)$ is true and $Q(x)$ is false and you show that this supposition leads to a contradiction. Direct proof, disproof by counterexample, and proof by contradiction can, therefore, all be viewed as three aspects of one whole. You arrive at one or the other by a thoughtful examination of the given statement, knowing what it means for a statement of that form to be true or false.

Another aim of the chapter is to help you obtain fundamental knowledge about numbers that is needed in mathematics and computer science. Note that the exercise sets contain problems of varying difficulty. Do not be discouraged if some of them are difficult for you.

Proofs given as solutions should be regarded as samples. Your instructor will probably discuss with you the particular range of proof styles that will be considered acceptable in your course.

Section 4.1

3. *a.* Yes, because

$$4rs = 2 \cdot (2rs)$$

and $2rs$ is an integer since r and s are integers and products of integers are integers.

b. Yes, because

$$6r + 4s^2 + 3 = 2(3r + 2s^2 + 1) + 1$$

and $3r + 2s^2 + 1$ is an integer since r and s are integers and products and sums of integers are integers.

c. Yes, because

$$r^2 + 2rs + s^2 = (r + s)^2$$

and $r + s$ is an integer that is greater than or equal to 2 since both r and s are positive integers and thus each is greater than or equal to 1.

6. For example, let $a = 1$ and $b = 0$. Then

$$\sqrt{a + b} = \sqrt{1} = 1$$

and also

$$\sqrt{a} + \sqrt{b} = \sqrt{1} + \sqrt{0} = 1.$$

Hence for these values of a and b,

$$\sqrt{a + b} = \sqrt{a} + \sqrt{b}.$$

In fact, if a is any nonzero integer and $b = 0$, then

$$\sqrt{a + b} = \sqrt{a + 0} = \sqrt{a} = \sqrt{a} + 0 = \sqrt{a} + \sqrt{0} = \sqrt{a} + \sqrt{b}.$$

12. <u>Counterexample:</u> Let $n = 5$. Then

$$\frac{n - 1}{2} = \frac{5 - 1}{2} = \frac{4}{2} = 2,$$

which is not odd.

24

15. According to the order of operations for real numbers, $-a^n = -(a^n)$. The following table shows that the property is true for some values of a and n and false for other values.

a	n	$-a^n$	$(-a)^n$	Does $-a^n = (-a)^n$?
0	2	$-0^2 = -0 = 0$	$(-0)^2 = 0^2 = 0$	Yes
3	2	$-3^2 = -(3^2) = -9$	$(-3)^2 = (-3)(-3) = 9$	No
-2	3	$-(-2)^3 = -((-2)^3) = -(-8) = 8$	$(-(-2))^3 = 2^3 = 8$	Yes
-3	2	$-(-3)^2 = -((-3)^2) = -9$	$(-(-3))^3 = 3^2 = 9$	No

18. $1^2 - 1 + 11 = 11$, which is prime. $2^2 - 2 + 11 = 13$, which is prime.
$3^2 - 3 + 11 = 17$, which is prime. $4^2 - 4 + 11 = 23$, which is prime.
$5^2 - 5 + 11 = 31$, which is prime. $6^2 - 6 + 11 = 41$, which is prime.
$7^2 - 7 + 11 = 53$, which is prime. $8^2 - 8 + 11 = 67$, which is prime.
$9^2 - 9 + 11 = 83$, which is prime. $10^2 - 10 + 11 = 101$, which is prime.

21. If a real number is greater than 1, then its square it greater than itself.

Start of Proof: Suppose x is any *[particular but arbitrarily chosen]* real number such that $x > 1$.

Conclusion to be shown: $x^2 > x$.

27. Proof 1:

Suppose m and n are any *[particular but arbitrarily chosen]* odd integers. *[We must show that $m + n$ is even.]*

By definition of odd, there exist integers r and s such that $m = 2r + 1$ and $n = 2s + 1$. Then

$$\begin{aligned} m + n &= (2r + 1) + (2s + 1) \quad \text{by substitution} \\ &= 2r + 2s + 2 \\ &= 2(r + s + 1) \quad\quad\quad \text{by algebra.} \end{aligned}$$

Let $u = r + s + 1$.

Then u is an integer because r, s, and 1 are integers and a sum of integers is an integer.

Hence $m + n = 2u$, where u is an integer, and so by definition of even, $m + n$ is even *[as was to be shown]*.

Proof 2:

Suppose m and n are any *[particular but arbitrarily chosen]* odd integers. *[We must show that $m + n$ is even.]*

By definition of odd, there exist integers r and s such that $m = 2r + 1$ and $n = 2s + 1$. Then

$$\begin{aligned} m + n &= (2r + 1) + (2s + 1) \quad \text{by substitution} \\ &= 2r + 2s + 2 \\ &= 2(r + s + 1) \quad\quad\quad \text{by algebra.} \end{aligned}$$

But $r + s + 1$ is an integer because r, s, and 1 are integers and a sum of integers is an integer.

Hence $m + n$ equals twice an integer, and so by definition of even, $m + n$ is even *[as was to be shown]*.

30. Proof:

Suppose m is any *[particular but arbitrarily chosen]* even integer. *[We must show that $3m + 5$ is odd.]*

By definition of even, $m = 2r$ for some integer r. Then

$$\begin{aligned} 3m + 5 &= 3(2r) + 5 & \text{by substitution} \\ &= 6r + 4 + 1 \\ &= 2(3r + 2) + 1 & \text{by algebra.} \end{aligned}$$

Let $t = 3r + 2$.

Then t is an integer because products and sums of integers are integers, and $3m + 5 = 2t + 1$.

Hence $3m + 5$ is odd by definition of odd *[as was to be shown]*.

36. To prove the given statement is false, we prove that its negation is true. The negation of the statement is "For all integers n, $6n^2 + 27$ is not prime."

    <u>Proof</u>:

    Suppose n is any integer. *[We must show that $6n^2 + 27$ is not prime.]*

    Note that $6n^2 + 27$ is positive because $n^2 \geq 0$ for all integers n and products and sums of positive real numbers are positive. Then

    $$6n^2 + 27 = 3(2n^2 + 9),$$

    and both 3 and $2n^2 + 9$ are positive integers each greater than 1 and less than $6n^2 + 27$.

    So $6n^2 + 27$ is not prime.

42. The mistake in the "proof" is that the same symbol, k, is used to represent two different quantities. By setting both m and n equal to $2k$, the "proof" specifies that $m = n$, and, therefore, it only deduces the conclusion in case $m = n$. If $m \neq n$, the conclusion is often false. For instance, $6 + 4 = 10$ but $10 \neq 4k$ for any integer k.

48. The statement is true.

    <u>Proof</u>:

    Let m and n be any even integers.

    By definition of even, $m = 2r$ and $n = 2s$ for some integers r and s.

    By substitution,

    $$m - n = 2r - 2s = 2(r - s).$$

    Since $r - s$ is an integer (being a difference of integers), then

    $m - n$ equals twice some integer, and so

    $m - n$ is even by definition of even.

51. The statement is false.

    <u>Counterexample</u>: Let $n = 2$. Then n is prime but

    $$(-1)^n = (-1)^2 = 1 \neq -1.$$

57. The statement is false.

    <u>Counterexample</u>: Let $m = n = 3$. Then $mn = 3 \cdot 3 = 9$, which is a perfect square, but neither m nor n is a perfect square.

60. The statement is false.

    <u>Counterexample</u>: When $a = 1$ and $b = 1$,

    $$\sqrt{a+b} = \sqrt{1+1} = \sqrt{2} \qquad \text{and} \qquad \sqrt{a} + \sqrt{b} = \sqrt{1} + \sqrt{1} = 2.$$

    But

    $$\sqrt{2} \neq 2, \qquad \text{and so} \qquad \sqrt{a+b} \neq \sqrt{a} + \sqrt{b}.$$

63. <u>Counterexample</u>: Let $n = 5$. Then

$$2^{2^n} + 1 = 2^{32} + 1 = 4,294,967,297 = (641) \cdot (6700417),$$

and so $2^{2^n} + 1$ is not prime.

Section 4.2

18. The statement is true.

<u>Proof</u>: Suppose r and s are any two distinct rational numbers. *[We must show that $\frac{r+s}{2}$ is rational.]*

By definition of rational, $r = \frac{a}{b}$ and $s = \frac{c}{d}$ for some integers a, b, c, and d with $b \neq 0$ and $d \neq 0$.

By substitution and the laws of algebra,

$$\frac{r+s}{2} = \frac{\frac{a}{b} + \frac{c}{d}}{2} = \frac{\frac{ad+bc}{bd}}{2} = \frac{ad+bc}{2bd}.$$

Now $ad + bc$ and $2bd$ are integers because a, b, c, and d are integers and products and sums of integers are integers. And $2bd \neq 0$ by the zero product property.

Hence $\frac{r+s}{2}$ is a quotient of integers with a nonzero denominator, and so $\frac{r+s}{2}$ is rational *[as was to be shown]*.

30. Let the quadratic equation be
$$x^2 + bx + c = 0$$

where b and c are rational numbers. Suppose one solution, r, is rational. Call the other solution s. Then
$$x^2 + bx + c = (x - r)(x - s) = x^2 - (r + s)x + rs.$$

By equating the coefficients of x,
$$b = -(r + s).$$

Solving for s yields
$$s = -r - b = -(r + b).$$

Because s is the negative of a sum of two rational numbers, s also is rational (by Theorem 4.2.2 and exercise 13).

33. *a.* Note that $(x - r)(x - s) = x^2 - (r + s)x + rs$.

If both r and s are odd integers, then $r + s$ is even and rs is odd (by properties 2 and 3).

If both r and s are even integers, then both $r + s$ and rs are even (by property 1).

If one of r and s is even and the other is odd, then $r + s$ is odd and rs is even (by properties 4 and 5).

b. It follows from part (a) that $x^2 - 1253x + 255$ cannot be written as a product of the form $(x - r)(x - s)$ because if it could be then $r + s$ would equal 1253 and rs would equal 255, both of which are odd integers. But for none of the possible cases (both r and s odd, both r and s even, and one of r and s odd and the other even) are both $r + s$ and rs odd integers.

Note: In Section 4.4, we establish formally that any integer is either even or odd. The type of reasoning used in this solution is called argument by contradiction. It is introduced formally in Section 4.6.

36. This incorrect proof just shows the theorem to be true in the one case where one of the rational numbers is 1/4 and the other is 1/2. It is an example of the mistake of arguing from examples. A correct proof must show the theorem is true for *any* two rational numbers.

39. This incorrect proof assumes what is to be proved. The second sentence asserts that a certain conclusion follows if $r + s$ is rational, and the rest of the proof uses that conclusion to deduce that $r + s$ is rational.

Section 4.3

3. Yes: $0 = 0 \cdot 5$.

9. Yes: $2a \cdot 34b = 4(17ab)$ and $17ab$ is an integer because a and b are integers and products of integers are integers.

21. The statement is true.

    Proof:

    Let m and n be any two even integers. *[We must show that mn is a multiple of 4.]*

    By definition of even, $m = 2r$ and $n = 2s$ for some integers r and s. Then

    $$
    \begin{aligned}
    mn &= (2r)(2s) & \text{by substitution} \\
    &= 4(rs) & \text{by algebra.}
    \end{aligned}
    $$

    Since rs is an integer (being a product of integers), mn is a multiple of 4 (by definition of divisibility).

27. The statement is false.

    Counterexample:

    Let $a = 2$, $b = 3$, and $c = 1$.

    Then $a \mid (b + c)$ because $b + c = 4$ and $2 \mid 4$ but $a \nmid b$ because $2 \nmid 3$ and $a \nmid c$ because $2 \nmid 1$.

30. The statement is false.

    Counterexample:

    Let $a = 4$ and $n = 6$.

    Then $a \mid n^2$ and $a \leq n$ because $4 \mid 36$ and $4 \leq 6$, but $a \nmid n$ because $4 \nmid 6$.

33. No. The values of nickels, dimes, and quarters are all multiples of 5. It follows from exercise 15 that a sum of numbers each of which is divisible by 5 is also divisible by 5. So since $4.72 is not a multiple of 5, $4.72 cannot be obtained using only nickels, dimes, and quarters.

    Note: The form of reasoning used in this answer is called argument by contradiction. It is discussed formally in Section 4.6.

36. *b.* Let $N = 12,858,306,120,312$.

    The sum of the digits of N is 42, which is divisible by 3 but not by 9. Therefore, N is divisible by 3 but not by 9.

    The right-most digit of N is neither 5 nor 0, and so N is not divisible by 5.

    The two right-most digits of N are 12, which is divisible by 4. Therefore, N is divisible by 4.

    c. Let $N = 517,924,440,926,512$.

    The sum of the digits of N is 61, which is not divisible by 3 (and hence not by 9 either). Therefore, N is not divisible by either 3 or 9.

The right-most digit of N is neither 5 nor 0, and so N is not divisible by 5.

The two right-most digits of N are 12, which is divisible by 4. Therefore, N is divisible by 4.

d. Let $N = 14,328,083,360,232$.

The sum of the digits of N is 45, which is divisible by 9 and hence also by 3. Therefore, N is divisible by 9 and by 3.

The right-most digit of N is neither 5 nor 0, and so N is not divisible by 5.

The two right-most digits of N are 32, which is divisible by 4. Therefore, N is divisible by 4.

39. *a.* $a^3 = p_1^{3e_1} \cdot p_2^{3e_2} \cdots p_k^{3e_k}$

b. To solve the problem, find the smallest powers of the factors 2, 3, 7, and 11 by which to multiply the given number so that the exponent of each factor becomes a multiple of 3. The value of k that works is $k = 2^2 \cdot 3 \cdot 7^2 \cdot 11$. Then

$$2^4 \cdot 3^5 \cdot 7 \cdot 11^2 \cdot k = 2^6 \cdot 3^6 \cdot 7^3 \cdot 11^3 = (2^2 \cdot 3^2 \cdot 7 \cdot 11)^3 = 2772^3.$$

42. *b.*
$$\begin{aligned}
20! &= 20 \cdot 19 \cdot 18 \cdot 17 \cdot 16 \cdot 15 \cdot 14 \cdot 13 \cdot 12 \cdot 11 \cdot 10 \cdot 9 \cdot 8 \cdot 7 \cdot 6 \cdot 5 \cdot 4 \cdot 3 \cdot 2 \cdot 1 \\
&= 2^2 \cdot 5 \cdot 19 \cdot 2 \cdot 3^2 \cdot 17 \cdot 2^4 \cdot 3 \cdot 5 \cdot 2 \cdot 7 \cdot 13 \cdot 2^2 \cdot 3 \cdot 11 \cdot 2 \cdot 5 \cdot 3^2 \cdot 2^3 \cdot 7 \cdot 2 \cdot 3 \cdot 5 \cdot 2^2 \cdot 3 \cdot 2 \\
&= 2^{18} \cdot 3^8 \cdot 5^4 \cdot 7^2 \cdot 11 \cdot 13 \cdot 17 \cdot 19
\end{aligned}$$

c. Squaring the result of part (b) gives

$$\begin{aligned}
(20!)^2 &= (2^{18} \cdot 3^8 \cdot 5^4 \cdot 7^2 \cdot 11 \cdot 13 \cdot 17 \cdot 19)^2 \\
&= 2^{36} \cdot 3^{16} \cdot 5^8 \cdot 7^4 \cdot 11^2 \cdot 13^2 \cdot 17^2 \cdot 19^2
\end{aligned}$$

When $(20!)^2$ is written in ordinary decimal form, there are as many zeros at the end of it as there are factors of the form $2 \cdot 5$ $(= 10)$ in its prime factorization.

Thus, since the prime factorization of $(20!)^2$ contains eight 5's and more than eight 2's, $(20!)^2$ contains eight factors of 10 and hence eight zeros.

45. <u>Proof:</u>

Suppose n is a nonnegative integer whose decimal representation ends in 5.

Then $n = 10m + 5$ for some integer m.

By factoring out a 5, $n = 10m + 5 = 5(2m + 1)$, and $2m + 1$ is an integer since m is an integer.

Hence n is divisible by 5 by definition of divisibility.

48. <u>Proof:</u>

Suppose n is any nonnegative integer for which the sum of the digits of n is divisible by 3.

By definition of decimal representation, n can be written in the form

$$n = d_k 10^k + d_{k-1} 10^{k-1} + \cdots + d_2 10^2 + d_1 10 + d_0$$

where k is a nonnegative integer and all the d_i are integers from 0 to 9 inclusive. Then

$$\begin{aligned}
n &= d_k (\underbrace{99\ldots9}_{k \text{ 9's}} + 1) + d_{k-1}(\underbrace{99\ldots9}_{(k-1) \text{ 9's}} + 1) + \cdots + d_2(99 + 1) + d_1(9 + 1) + d_0 \\
&= d_k \cdot \underbrace{99\ldots9}_{k \text{ 9's}} + d_{k-1} \cdot \underbrace{99\ldots9}_{(k-1) \text{ 9's}} + \cdots + d_2 \cdot 99 + d_1 \cdot 9 + (d_k + d_{k-1} + \cdots + d_2 + d_1 + d_0) \\
&= 9(d_k \cdot \underbrace{11\ldots1}_{k \text{ 1's}} + d_{k-1} \cdot \underbrace{11\ldots1}_{(k-1) \text{ 1's}} + \cdots + d_2 \cdot 11 + d_1) + (d_k + d_{k-1} + \cdots + d_2 + d_1 + d_0) \\
&= 3[3(d_k \cdot \underbrace{11\ldots1}_{k \text{ 1's}} + d_{k-1} \cdot \underbrace{11\ldots1}_{(k-1) \text{ 1's}} + \cdots + d_2 \cdot 11 + d_1)] + (d_k + d_{k-1} + \cdots + d_2 + d_1 + d_0) \\
&= (\text{an integer divisible by 3}) + (\text{the sum of the digits of } n).
\end{aligned}$$

Since the sum of the digits of n is divisible by 3, n can be written as a sum of two integers each of which is divisible by 3.

It follows from exercise 15 that n is divisible by 3.

Section 4.4

6. $q = -4$, $r = 5$

9. *a.* 5 *b.* 3

12. Let the days of the week be numbered from 0 (Sunday) through 6 (Saturday) and let $DayT$ and $DayN$ be variables representing the day of the week today and the day of the week N days from today. By the quotient-remainder theorem, there exist unique integers q and r such that

$$DayT + N = 7q + r \text{ and } 0 \leq r < 7.$$

Now $DayT + N$ counts the number of days to the day N days from today starting last Sunday (where "last Sunday" is interpreted to mean today if today is a Sunday). Thus $DayN$ is the day of the week that is $DayT + N$ days from last Sunday. Because each week has seven days,

$DayN$ is the same as the day of the week $DayT + N - 7q$ days from last Sunday.

But

$$DayT + N - 7q = r \quad \text{and} \quad 0 \leq r < 7.$$

Therefore,

$$DayN = r = (DayT + N) \bmod 7.$$

15. There are 13 leap year days between January 1, 2000 and January 1, 2050 (once every four years in 2000, 2004, 2008, 2012, . . . , 2048). So 13 of the years have 366 days and the remaining 38 years have 365 days. This gives a total of

$$13 \cdot 366 + 37 \cdot 365 = 18,263$$

days between the two dates. Using the formula $DayN = (DayT + N) \bmod 7$, and letting $DayT = 6$ (Saturday) and $N = 18,263$ gives

$$DayN = (6 + 18263) \bmod 7 = 18269 \bmod 7 = 6,$$

which is also a Saturday.

21. Given that b is an integer and $b \bmod 12 = 5$, it follows that $8b \bmod 12 = 4$.

    <u>Proof:</u> When b is divided by 12, the remainder is 5. Thus there exists an integer m so that $b = 12m + 5$. Multiplying this equation by 8 gives

$$
\begin{aligned}
8b &= 8(12m + 5) && \text{by substitution} \\
&= 96m + 40 \\
&= 96m + 36 + 4 \\
&= 12(8m + 3) + 4 && \text{by algebra.}
\end{aligned}
$$

Since $8m + 3$ is an integer and since $0 \leq 4 < 12$, the uniqueness part of the quotient-remainder theorem guarantees that the remainder obtained when $8b$ is divided by 12 is 4.

24. <u>Proof:</u>

    Suppose m and n are any *[particular but arbitrarily chosen]* integers such that $m \bmod 5 = 2$ and $n \bmod 5 = 1$.

Then the remainder obtained when m is divided by 5 is 2 and the remainder obtained when n is divided by 5 is 1, and so $m = 5q + 2$ and $n = 5r + 1$ for some integers q and r.

$$
\begin{aligned}
mn &= (5q + 2)(5r + 1) && \text{by substitution} \\
&= 25qr + 5q + 10r + 2 \\
&= 5(5qr + q + 2r) + 2 && \text{by algebra.}
\end{aligned}
$$

Because products and sums of integers are integers, $5qr + q + 2r$ is an integer, and hence $mn = 5 \cdot (\text{an integer}) + 2$.

Thus, since $0 \le 2 < 5$, the remainder obtained when mn is divided by 5 is 2, and so $mn \bmod 5 = 2$.

30. a. <u>Proof</u>: Suppose n and $n + 1$ are any two consecutive integers. By the quotient-remainder theorem with $d = 3$, we know that $n = 3q$, or $n = 3q + 1$, or $n = 3q + 2$ for some integer q.

Case 1 ($n = 3q$ for some integer q): In this case,

$$
\begin{aligned}
n(n + 1) &= 3q(3q + 1) && \text{by substitution} \\
&= 3[q(3q + 1)] && \text{by algebra.}
\end{aligned}
$$

Let $k = q(3q + 1)$. Then k is an integer because sums and products of integers are integers. Hence $n(n + 1) = 3k$ for some integer k.

Case 2 ($n = 3q + 1$ for some integer q): In this case,

$$
\begin{aligned}
n(n + 1) &= (3q + 1)(3q + 2) && \text{by substitution} \\
&= 9q^2 + 9q + 2 \\
&= 3(3q^2 + 3q) + 2 && \text{by algebra.}
\end{aligned}
$$

Let $k = 3q^2 + 3q$. Then k is an integer because sums and products of integers are integers. Hence $n(n + 1) = 3k + 2$ for some integer k.

Case 3 ($n = 3q$ for some integer q): In this case,

$$
\begin{aligned}
n(n + 1) &= (3q + 2)(3q + 3) && \text{by substitution} \\
&= 3[(3q + 2)(q + 1)] \\
&= 3[(3q + 2)(q + 1)] && \text{by algebra.}
\end{aligned}
$$

Let $k = (3q + 2)(q + 1)$. Then k is an integer because sums and products of integers are integers. Hence $n(n + 1) = 3k$ for some integer k.

Conclusion: In all three cases, the product of the two consecutive integers either equals $3k$ or it equals $3k + 2$ for some integer k [*as was to be shown*].

b. Given any integers m and n, $mn \bmod 3 = 0$ or $mn \bmod 3 = 2$.

(Or, equivalently, Given any integer n, $mn \bmod 3 \ne 1$.)

33. Given any integers a, b, and c if $a - b$ is odd and $b - c$ is even, then $a - c$ is odd.

<u>Proof</u>: Suppose a, b, and c are any integers such that $a - b$ is odd and $b - c$ is even. Then $(a - b) + (b - c)$ is a sum of an odd integer and an even integer and hence is odd (by property 5 in Example 4.2.3). But $(a - b) + (b - c) = a - c$, and thus $a - c$ is odd.

36. <u>Proof</u>:

Suppose n is any integer. [*We must show that* $8 \mid n(n + 1)(n + 2)(n + 3)$.]

By the quotient-remainder theorem with $d = 4$, there is an integer k such that

$$
n = 4k \quad \text{or} \quad n = 4k + 1 \quad \text{or} \quad n = 4k + 2 \quad \text{or} \quad n = 4k + 3.
$$

Case 1 (n = 4k for some integer k): In this case,

$$\begin{aligned} n(n+1)(n+2)(n+3) &= 4k(4k+1)(4k+2)(4k+3) \quad \text{by substitution} \\ &= 8[k(4k+1)(2k+1)(4k+3)] \quad \text{by algebra,} \end{aligned}$$

and this is divisible by 8 (because k is an integer and sums and products of integers are integers).

Case 2 (n = 4k + 1 for some integer k): In this case,

$$\begin{aligned} n(n+1)(n+2)(n+3) &= (4k+1)(4k+2)(4k+3)(4k+4) \quad \text{by substitution} \\ &= 8[(4k+1)(2k+1)(4k+3)(k+1)] \quad \text{by algebra,} \end{aligned}$$

and this is divisible by 8 (because k is an integer and sums and products of integers are integers).

Case 3 (n = 4k + 2 for some integer k): In this case,

$$\begin{aligned} n(n+1)(n+2)(n+3) &= (4k+2)(4k+3)(4k+4)(4k+5) \quad \text{by substitution} \\ &= 8[(2k+1)(4k+3)(k+1)(4k+5)] \quad \text{by algebra,} \end{aligned}$$

and this is divisible by 8 (because k is an integer and sums and products of integers are integers).

Case 4 (n = 4k + 3 for some integer k): In this case,

$$\begin{aligned} n(n+1)(n+2)(n+3) &= (4k+3)(4k+4)(4k+5)(4k+6) \quad \text{by substitution} \\ &= 8[(4k+3)(k+1)(4k+5)(2k+3)] \quad \text{by algebra,} \end{aligned}$$

and this is divisible by 8 (because k is an integer and sums and products of integers are integers).

Conclusion: In all four possible cases, $8 \mid n(n+1)(n+2)(n+3)$ *[as was to be shown]*.

Note: One can make use of exercise 17 to produce a proof that only requires two cases: n is even and n is odd. Then, since both $n(n+1)$ and $(n+2)(n+3)$ are products of consecutive integers, by the result of exercise 17, both products are even and hence contain a factor of 2. Multiplying those two factors shows that there is a factor of 4 in $n(n+1)(n+2)(n+3)$.

39. Proof: Consider any four consecutive integers. Call the smallest n. Then the sum of the four integers is

$$n + (n+1) + (n+2) + (n+3) = 4n + 6 = 4(n+1) + 2.$$

Let $k = n + 1$. Then k is an integer because it is a sum of integers. Hence n can be written in the required form.

42. Proof: Let p be any prime number except 2 or 3. By the quotient-remainder theorem, there is an integer k so that p can be written as

$$p = 6k \quad \text{or} \quad p = 6k+1 \quad \text{or} \quad p = 6k+2 \quad \text{or} \quad p = 6k+3 \quad \text{or} \quad p = 6k+4 \quad \text{or} \quad p = 6k+5.$$

Since p is prime and $p \neq 2$, p is not divisible by 2. Consequently, $p \neq 6k$, $p \neq 6k+2$, and $p \neq 6k+4$ for any integer k *[because all of these numbers are divisible by 2]*.

Furthermore, since p is prime and $p \neq 3$, p is not divisible by 3. Thus $p \neq 6k+3$ *[because this number is divisible by 3]*.

Therefore, $p = 6k+1$ or $p = 6k+5$ for some integer k.

45. <u>Proof</u>: Let c be any positive real number and let r be any real number. Suppose that $-c \leq r \leq c$.(*) By the trichotomy law (see Appendix A, T17), either $r \geq 0$ or $r < 0$.

    Case 1 ($r \geq 0$): In this case $|r| = r$, and so by substitution into (*), $-c \leq |r| \leq c$. In particular, $|r| \leq c$.

    Case 2 ($r < 0$): In this case $|r| = -r$, and so $r = -|r|$. Hence by substitution into (*), $-c \leq -|r| \leq c$. In particular, $-c \leq -|r|$. Multiplying both sides by -1 gives $c \geq |r|$, or, equivalently, $|r| \leq c$.

    Therefore, regardless of whether $r \geq 0$ or $r < 0$, $|r| \leq c$ *[as was to be shown]*.

48. *Solution 1*: We are given that M is a matrix with m rows and n columns, stored in row major form at locations $N + k$, where $0 \leq k < mn$. Given a value for k, we want to find indices r and s so that the entry for M in row r and column s, a_{rs}, is stored in location $N + k$. By the quotient-remainder theorem, $k = nQ + R$, where Q and R are integers and $0 \leq R < n$. The first Q rows of M (each of length n) are stored in the first nQ locations: $N + 0, N + 1, \ldots, N + nQ - 1$ with a_{Qn} stored in the last of these. Consider the next row. When $r = Q + 1$,

    $$a_{r1} \text{ will be in location } N + nQ$$
    $$a_{r2} \text{ will be in location } N + nQ + 1$$
    $$a_{r3} \text{ will be in location } N + nQ + 2$$
    $$\vdots$$
    $$a_{rs} \text{ will be in location } N + nQ + (s - 1)$$
    $$\vdots$$

    and a_{rn} will be in location $N + nQ + (n - 1) = N + n(Q + 1) - 1$.

    Thus location $N + k$ contains a_{rs} where $r = Q + 1$ and $R = s - 1$. But $Q = k \ div \ n$ and $R = k \ mod \ n$, and hence $r = (k \ div \ n) + 1$ and $s = (k \ mod \ n) + 1$.

    Solution 2: *[After the floor notation has been introduced, the following solution can be considered as an alternative.]* To find a formula for r, note that for $1 \leq a \leq m$,

    $$\text{when} \quad (a - 1)n \leq k < an, \quad \text{then} \quad r = a.$$

    Dividing through by n gives that

    $$\text{when} \quad (a - 1) \leq \frac{k}{n} < a, \quad \text{then} \quad r = a.$$

    or, equivalently,

    $$\text{when} \quad \left\lfloor \frac{k}{n} \right\rfloor = a - 1, \quad \text{then} \quad r = a.$$

    But this implies that

    $$\text{when} \quad a = \left\lfloor \frac{k}{n} \right\rfloor + 1 = r, \quad \text{then} \quad r = a.$$

    and since

    $$\left\lfloor \frac{k}{n} \right\rfloor = k \ div \ n, \quad \text{we have that} \quad r = (k \ div \ n) + 1.$$

    To find a formula for s, note that

    $$\text{when} \quad k = n \cdot (\text{an integer}) + b \quad \text{and} \quad 0 \leq b < n, \quad \text{then} \quad s = b + 1.$$

    Thus by the quotient-remainder theorem, $s = (k \ mod \ n) + 1$.

51. Answer to the first question: not necessarily

Counterexample: Let $m = n = 3$, $d = 2$, $a = 1$, and $b = 1$. Then

$$m \bmod d = n \bmod d = 3 \bmod 2 = 1 = a = b.$$

But $a + b = 1 + 1 = 2$, whereas $(m + n) \bmod d = 6 \bmod 2 = 0$.

Answer to the second question: yes.

Proof: Suppose m, n, a, b, and d are integers and $m \bmod d = a$ and $n \bmod d = b$. By definition of mod, $m = dq_1 + a$ and $n = dq_2 + b$ for some integers q_1 and q_2. Then

$$
\begin{aligned}
m + n &= (dq_1 + a) + (dq_2 + b) && \text{by substitution} \\
&= d(q_1 + q_2) + (a + b) \ (\text{*}) && \text{by algebra.}
\end{aligned}
$$

Apply the quotient-remainder theorem to $a + b$ to obtain unique integers q_3 and r such that $a + b = dq_3 + r$ (**) and $0 \le r < d$. By definition of mod, $r = (a + b) \bmod d$. Then

$$
\begin{aligned}
m + n &= d(q_1 + q_2) + (a + b) && \text{by (*)} \\
&= d(q_1 + q_2) + (dq_3 + r) && \text{by (**)} \\
&= d(q_1 + q_2 + q_3) + r && \text{by algebra,}
\end{aligned}
$$

where $q_1 + q_2 + q_3$ and r are integers and $0 \le r < d$. Hence by definition of mod,

$$r = (m + n) \bmod d, \text{ and so } (m + n) \bmod d = (a + b) \bmod d.$$

Section 4.5

6. If k is an integer, then $\lceil k \rceil = k$ because $k - 1 < k \le k$ and $k - 1$ and k are integers.

9. If the remainder obtained when n is divided by 36 is positive, an additional box beyond those containing exactly 36 units will be needed to hold the extra units.

So, since the ceiling notation rounds each number up to the nearest integer, the number of boxes required will be $\lceil \frac{n}{36} \rceil$.

Also, because the ceiling of an integer is itself, if the number of units is a multiple of 36, the number of boxes required will be $\lceil \frac{n}{36} \rceil$ as well.

Thus the ceiling notation is more appropriate for this problem because the answer is simply $\lceil \frac{n}{36} \rceil$ regardless of the value of n.

If the floor notation is used, the answer is more complicated: if $\frac{n}{36}$ is not an integer, the answer is $\lfloor \frac{n}{36} \rfloor + 1$, but if n is an integer, it is $\lfloor \frac{n}{36} \rfloor$.

18. Counterexample: Let $x = y = 1.5$. Then

$$\lceil x + y \rceil = \lceil 1.5 + 1.5 \rceil = \lceil 3 \rceil = 3 \quad \text{whereas} \quad \lceil x \rceil + \lceil y \rceil = \lceil 1.5 \rceil + \lceil 1.5 \rceil = 2 + 2 = 4,$$

and $3 \ne 4$.

21. Proof: Let n be any odd integer. *[We must show that $\lceil \frac{n}{2} \rceil = \frac{n+1}{2}$.]* By definition of odd, $n = 2k + 1$ for some integer k. The left-hand side of the equation to be proved is

$$
\begin{aligned}
\left\lceil \frac{n}{2} \right\rceil &= \left\lceil \frac{2k + 1}{2} \right\rceil && \text{by substitution} \\[2mm]
&= \left\lceil k + \frac{1}{2} \right\rceil && \text{by algebra} \\[2mm]
&= k + 1 && \text{by definition of ceiling because } k \text{ is} \\
& && \text{an integer and } k < k + 1/2 \le k + 1
\end{aligned}
$$

On the other hand, the right-hand side of the equation to be proved is

$$\frac{n+1}{2} = \frac{(2k+1)+1}{2} \qquad \text{by substitution}$$

$$= \frac{2k+2}{2}$$

$$= \frac{2(k+1)}{2}$$

$$= k+1 \qquad \text{by algebra.}$$

Thus both the left- and right-hand sides of the equation to be proved equal $k+1$, and so both are equal to each other. In other words, $\lceil n/2 \rceil = (n+1)/2$ *[as was to be shown]*.

24. <u>Proof</u>:

Suppose m is any integer and x is any real number that is not an integer.

By definition of floor, $\lfloor x \rfloor = n$ where n is an integer and $n \leq x < n+1$.

Since x is not an integer, $x \neq n$, and so

$$n < x < n+1.$$

Multiply all parts of this inequality by -1 to obtain

$$-n > -x > -n-1.$$

Then add m to all parts to obtain

$$m-n > m-x > m-n-1, \quad \text{or, equivalently,} \quad m-n-1 < m-x < m-n.$$

But $m-n-1$ and $m-n$ are both integers, and so by definition of floor, $\lfloor m-x \rfloor = m-n-1$. By substitution,

$$\lfloor x \rfloor + \lfloor m-x \rfloor = n + (m-n-1) = m-1$$

[as was to be shown].

27. <u>Proof</u>: Suppose x is any real number such that

$$x - \lfloor x \rfloor \geq 1/2.$$

Multiply both sides by 2 to obtain

$$2x - 2\lfloor x \rfloor \geq 1 \quad \text{or, equivalently,} \quad 2x \geq 2\lfloor x \rfloor + 1.$$

Now by definition of floor,

$$x < \lfloor x \rfloor + 1 \quad \text{and hence} \quad 2x < 2\lfloor x \rfloor + 2.$$

Put the two inequalities involving x together to obtain

$$2\lfloor x \rfloor + 1 \leq 2x < 2\lfloor x \rfloor + 2.$$

By definition of floor, then, $\lfloor 2x \rfloor = 2\lfloor x \rfloor + 1$.

Section 4.6

6. *Negation of Statement*: There is a greatest negative real number.

   <u>Proof of statement (by contradiction)</u>:

   Suppose not. That is, suppose there is a greatest negative real number. Call it a. *[We must show that this supposition leads logically to a contradiction.]*

   By supposition,

   $$a < 0 \quad \text{and} \quad a \geq x \quad \text{for every negative real number } x.$$

   Let $b = a/2$. Then b is a real number because b is a quotient of two real numbers (with a nonzero denominator). Also

   $$\text{because} \quad 0 < \frac{1}{2} < 1 \quad \text{then} \quad 0 > \frac{a}{2} > a$$

   by multiplying all parts of the inequality by a, which is negative. Thus

   $$a < \frac{a}{2} < 0 \quad \text{and so, by substitution,} \quad a < b < 0.$$

   Thus b is a negative real number that is greater than a. This contradicts the supposition that a is the greatest negative real number.

   [Hence the supposition is false and the given statement is true.]

9. *b*. <u>Proof by contradiction</u>:

   Suppose not. That is, suppose there is an irrational number and a rational number whose difference is rational. In other words, suppose there are real numbers x and y such that x is irrational, y is rational and $x - y$ is rational. *[We must show that this supposition leads logically to a contradiction.]*

   By definition of rational,

   $$y = \frac{a}{b} \quad \text{and} \quad x - y = \frac{c}{d} \quad \text{for some integers } a, b, c, \text{ and } d \text{ with } b \neq 0 \text{ and } d \neq 0.$$

   Then, by substitution,

   $$x - \frac{a}{b} = \frac{c}{d}.$$

   Solve this equation for x to obtain

   $$x = \frac{c}{d} + \frac{a}{b} = \frac{bc}{bd} + \frac{ad}{bd} = \frac{bc + ad}{bd}.$$

   But both $bc + ad$ and bd are integers because products and sums of integers are integers, and $bd \neq 0$ by the zero product property.

   Hence x is a ratio of integers with a nonzero denominator, and so x is rational by definition of rational.

   This contradicts the supposition that x is irrational. *[Hence the supposition is false and the given statement is true.]*

   Note: The fact that order matters in subtraction implies that the truth of the statement in exercise 8 does not automatically imply the truth of the statement in this exercise.

12. Proof by contradiction:

Suppose not. That is, suppose there are rational numbers a and b such that $b \neq 0$, r is an irrational number, and $a + br$ is rational. *[We must show that this supposition leads logically to a contradiction.]*

By definition of rational,

$$a = \frac{i}{j}, \quad b = \frac{k}{l} \quad \text{and} \quad a + br = \frac{m}{n}$$

where i, j, k, l, m, and n are integers and $j \neq 0$, $l \neq 0$, and $n \neq 0$. Since $b \neq 0$, we also have that $k \neq 0$. By substitution

$$a + br = \frac{i}{j} + \frac{k}{l} \cdot r = \frac{m}{n}, \quad \text{or, equivalently,} \quad \frac{k}{l} \cdot r = \frac{m}{n} - \frac{i}{j}$$

Solving for r gives

$$r = \frac{mj - in}{nj} \cdot \frac{l}{k} = \frac{mjl - inl}{njk}.$$

Now $mjl - inl$ and njk are both integers *[because products and differences of integers are integers]* and $njk \neq 0$ because $n \neq 0, j \neq 0$, and $k \neq 0$.

It follows, by definition of rational, that r is a rational number, which contradicts the supposition that r is irrational. *[Hence the supposition is false and the given statement is true.]*

15. Yes.

Proof by contradiction:

Suppose not. That is, suppose there exist integers a, b, and c such that a and b are both odd and $a^2 + b^2 = c^2$. *[We must show that this supposition leads logically to a contradiction.]*

By definition of odd, $a = 2k + 1$ and $b = 2m + 1$ for some integers k and m. Then,

$$
\begin{aligned}
c^2 &= a^2 + b^2 && \text{by supposition} \\
&= (2k + 1)^2 + (2m + 1)^2 && \text{by substitution} \\
&= 4k^2 + 4k + 1 + 4m^2 + 4m + 1 \\
&= 4(k^2 + k + m^2 + m) + 2 && \text{by algebra.}
\end{aligned}
$$

Let $t = k^2 + k + m^2 + m$. Then t is an integer because products and sums of integers are integers, and so, by substitution and algebra,

$$c^2 = 4t + 2 = 2(2t + 1).$$

Thus c^2 is even by definition of even, and hence c is even by Proposition 4.6.4. It follows by definition of even again that $c = 2r$ for some integer r. Substituting $c = 2r$ into the equation $c^2 = 4t + 2$ gives

$$(2r)^2 = 4t + 2, \quad \text{and so} \quad 4r^2 = 4t + 2.$$

Dividing by 2 gives

$$2r^2 = 2t + 1.$$

But since r^2 is an integer, $2r^2$ is even, by definition of even, and hence $2t + 1$ is even. However, since t is an integer, $2t + 1$ is odd by definition of odd. So $2t + 1$ is both even and odd, which contradicts Theorem 4.6.2 . It follows that the supposition that there exist integers a, b, and c such that a and b are both odd and $a^2 + b^2 = c^2$ is false, and thus the given statement is true.

24. *a. Proof by contraposition:*

Suppose x is a nonzero real number and the reciprocal of x, namely $\frac{1}{x}$, is rational. *[We must show that x is rational.]*

Because $\frac{1}{x}$ is rational, there are integers a and b with $b \neq 0$ such that

$$\frac{1}{x} = \frac{a}{b} \ (*).$$

Now

since $1 \cdot \left(\frac{1}{x}\right) = 1$, it follows that $\frac{1}{x}$ cannot equal zero, and so $a \neq 0$.

Thus we may solve equation (*) for x to obtain

$$x = \frac{b}{a} \quad \text{where } b \text{ and } a \text{ are integers and } a \neq 0.$$

Hence, by definition of rational, $\frac{1}{x}$ is rational *[as was to be shown]*.

b. Proof by contradiction:

Suppose not. That is, suppose there exists a nonzero irrational number x such that the reciprocal of x, namely $\frac{1}{x}$, is rational. *[We must show that this supposition leads logically to a contradiction.]*

By definition of rational,

$$\frac{1}{x} = \frac{a}{b} \ (*).$$

where a and b are integers with $b \neq 0$. Now

since $1 \cdot \left(\frac{1}{x}\right) = 1$, it follows that $\frac{1}{x}$ cannot equal zero, and so $a \neq 0$.

Thus we may solve equation (*) for x to obtain

$$x = \frac{b}{a} \quad \text{where } b \text{ and } a \text{ are integers and } a \neq 0.$$

Hence, by definition of rational, $\frac{1}{x}$ is rational, which contradicts the supposition that x is irrational. *[Hence the supposition is false and the given statement is true.]*

27. *a.* Proof 1 by contraposition: *[This proof is based only on the definitions of even and odd integers and the parity property.]*

Suppose m and n are integers such that one of m and n is even and the other is odd. *[We must show that $m + n$ is not even.]*

Case 1 (m is even and n is odd): In this case there exists integers r and s such that $m = 2r$ and $n = 2s + 1$. Then

$$\begin{aligned} m + n &= 2r + (2s + 1) && \text{by substitution} \\ &= 2(r + s) + 1 && \text{by algebra.} \end{aligned}$$

Let $k = r + s$. Then k is an integer because it is a product of integers, and thus $m + n$ is odd by definition of odd.

Case 2 (m is odd and n is even): In this case, by Theorem 4.4.3, there is an integer m such that $n^2 = 8m + 1$. Then

$$\begin{aligned} m + n &= (2r + 1) + 2s && \text{by substitution} \\ &= 2(r + s) + 1 && \text{by algebra.} \end{aligned}$$

Let $k = r + s$. Then k is an integer because it is a product of integers, and thus $m + n$ is odd by definition of odd.

Conclusion: In both cases, $m + n$ is odd, and so by the parity property, $m + n$ is not even *[as was to be shown]*.

<u>Proof 2 by contraposition</u>: *[This proof uses a previously established property of even and odd integers.]*

Suppose m and n are integers such that one of m and n is even and the other is odd. *[We must show that $m + n$ is not even.]*

By property 5 of Example 4.2.3, the sum of any even integer and any odd integer is odd. Hence $m + n$ is odd, and so, by the parity property, $m + n$ is not even *[as was to be shown]*.

b. <u>Proof 1 by contradiction</u>: *[This proof is based only on the definitions of even and odd integers and the parity property.]*

Suppose not. That is, suppose there exist integers m and n such that $m + n$ is even and either m is even and n is odd or m is odd and n is even. *[We must show that this supposition leads logically to a contradiction.]*

Case 1 (m is even and n is odd): In this case there exists integers r and s such that $m = 2r$ and $n = 2s + 1$. Then

$$
\begin{aligned}
m + n &= 2r + (2s + 1) & \text{by substitution} \\
&= 2(r + s) + 1 & \text{by algebra.}
\end{aligned}
$$

Let $k = r + s$. Then k is an integer because it is a product of integers, and thus $m + n$ is odd by definition of odd.

Case 2 (m is odd and n is even): In this case, by Theorem 4.4.3, there is an integer m such that $n^2 = 8m + 1$. Then

$$
\begin{aligned}
m + n &= (2r + 1) + 2s & \text{by substitution} \\
&= 2(r + s) + 1 & \text{by algebra.}
\end{aligned}
$$

Let $k = r + s$. Then k is an integer because it is a product of integers, and thus $m + n$ is odd by definition of odd.

Conclusion: In both cases, $m + n$ is odd, whereas, by supposition, $m + n$ is even. This result contradicts the parity property, which says that an integer cannot be both even and odd. Thus the supposition is false, and the given statement is true.

<u>Proof 2 by contradiction</u>: *[This proof uses a previously established property of even and odd integers.]*

Suppose not. That is, suppose there exist integers m and n such that $m + n$ is even and either m is even and n is odd or m is odd and n is even. *[We must show that this supposition leads logically to a contradiction.]*

By property 5 of Example 4.2.3, the sum of any even integer and any odd integer is odd. Thus both when m is even and n is odd and when m is odd and n is even, the sum $m + n$ is odd. But, by supposition, $m + n$ is even. This result contradicts the parity property, which says that an integer cannot be both even and odd. Thus the supposition is false, and the given statement is true.

33. After crossing out all multiples of 2, 3, 5, and 7 (the prime numbers less than $\sqrt{100}$), the remaining numbers are prime. They are circled in the following diagram.

(2) (3) ~~4~~ (5) ~~6~~ (7) ~~8~~ ~~9~~ ~~10~~ (11) ~~12~~ (13) ~~14~~ ~~15~~

~~16~~ (17) ~~18~~ (19) ~~20~~ ~~21~~ ~~22~~ (23) ~~24~~ ~~25~~ ~~26~~ ~~27~~ ~~28~~ (29)

~~30~~ (31) ~~32~~ ~~33~~ ~~34~~ ~~35~~ ~~36~~ (37) ~~38~~ ~~39~~ ~~40~~ (41) ~~42~~ (43)

~~44~~ ~~45~~ ~~46~~ (47) ~~48~~ ~~49~~ ~~50~~ ~~51~~ ~~52~~ (53) ~~54~~ ~~55~~ ~~56~~ ~~57~~

~~58~~ (59) ~~60~~ (61) ~~62~~ ~~63~~ ~~64~~ ~~65~~ ~~66~~ (67) ~~68~~ ~~69~~ ~~70~~ (71)

~~72~~ (73) ~~74~~ ~~75~~ ~~76~~ ~~77~~ ~~78~~ (79) ~~80~~ ~~81~~ ~~82~~ (83) ~~84~~ ~~85~~

~~86~~ ~~87~~ ~~88~~ (89) ~~90~~ ~~91~~ ~~92~~ ~~93~~ ~~94~~ ~~95~~ ~~96~~ (97) ~~98~~ ~~99~~

Section 4.7

6. False.

Proof 1 (by using a previous result): $\sqrt{2}/6 = (1/6) \cdot \sqrt{2}$, which is a product of a nonzero rational number and an irrational number. By exercise 11 of Section 4.6, such a product is irrational.

Proof 2 (by contradiction):

Suppose not. That is, suppose $\sqrt{2}/6$ is rational. *[We must show that this supposition leads logically to a contradiction.]*

By definition of rational, there exist integers a and b with

$$\sqrt{2}/6 = a/b \quad \text{and} \quad b \neq 0.$$

Solving for $\sqrt{2}$ gives $\sqrt{2} = 6a/b$. But $6a$ is an integer (because products of integers are integers) and b is a nonzero integer.

Therefore, by definition of rational, $\sqrt{2}$ is rational. This contradicts Theorem 4.7.1 which states that $\sqrt{2}$ is irrational. Hence the supposition is false. In other words $\sqrt{2}/6$ is irrational.

12. *Counterexample:* $\sqrt{2}$ is irrational. Also $\sqrt{2} \cdot \sqrt{2} = 2$ and 2 is rational because $2 = 2/1$. Thus there exist irrational numbers whose product is rational.

15. *a. Proof by contraposition:*

Let n be any integer such that n is not even. *[We must show that n^3 is not even.]*

By Theorem 4.6.2 n is odd, and so n^3 is also odd (by property 3 of Example 4.2.3 applied twice).

Thus (again by Theorem 4.6.2), n^3 is not even *[as was to be shown]*.

b. Proof by contradiction:

Suppose not. That is, suppose $\sqrt[3]{2}$ is rational. *[We must show that this supposition leads logically to a contradiction.]*

By definition of rational,

$$\sqrt[3]{2} = a/b \text{ for some integers } a \text{ and } b \text{ with } b \neq 0.$$

By cancelling any common factors if necessary, we may assume that a and b have no common factors. Cubing both sides of equation $\sqrt[3]{2} = a/b$ gives

$$2 = a^3/b^3, \quad \text{and so} \quad 2b^3 = a^3.$$

Thus a^3 is even. By part (a) of this question, a is even, and thus $a = 2k$ for some integer k. By substitution

$$a^3 = (2k)^3 = 8k^3 = 2b^3,$$

and so

$$b^3 = 4k^3 = 2(2k^3).$$

It follows that b^3 is even, and hence (also by part (a)) b is even.

Thus both a and b are even which contradicts the assumption that a and b have no common factor. Therefore, the supposition is false, and so $\sqrt[3]{2}$ is irrational.

18. <u>Proof</u> :

Suppose that a and d are integers with $d > 0$ and that q_1, q_2, r_1, and r_2 are integers such that

$$a = dq_1 + r_1 \quad \text{and} \quad a = dq_2 + r_2, \quad \text{where} \quad 0 \le r_1 < d \quad \text{and} \quad 0 \le r_2 < d.$$

[We must show that $r_1 = r_2$ and $q_1 = q_2$.]

Then
$$dq_1 + r_1 = dq_2 + r_2,$$

and so
$$r_2 - r_1 = dq_1 - dq_2 = d(q_1 - q_2).$$

This implies that
$$d \mid (r_2 - r_1)$$

because $q_1 - q_2$ is an integer (since it is a difference of integers).

But both r_1 and r_2 lie between 0 and d, and thus the difference $r_2 - r_1$ lies between $-d$ and d.

[For, by properties T23 and T26 of Appendix A, because $0 \le r_1 < d$ and $0 \le r_2 < d$, then multiplying the first inequality by -1 gives $0 \ge -r_1 > -d$ or, equivalently, $-d < -r_1 \le 0$, and adding the inequalities $-d < -r_1 \le 0$ and $0 \le r_2 < d$ gives $-d < r_2 - r_1 < d$.]

Since $r_2 - r_1$ is a multiple of d and yet lies between $-d$ and d, the only possibility is that $r_2 - r_1 = 0$, or, equivalently, that $r_1 = r_2$.

Substituting back into the original expressions for a and equating the two gives

$$dq_1 + r_1 = dq_2 + r_1 \qquad \text{because } r_1 = r_2.$$

Subtracting r_1 from both sides gives $dq_1 = dq_2$, and since $d \ne 0$, we have that $q_1 = q_2$.

24. <u>Proof by contradiction</u> :

Suppose not. That is, suppose that $\log_5(2)$ is rational. *[We will show that this supposition leads logically to a contradiction.]*

By definition of rational,

$$\log_5(2) = \frac{a}{b} \quad \text{for some integers} \quad a \text{ and } b \text{ with } b \ne 0.$$

Since logarithms are always positive, we may assume that a and b are both positive. By definition of logarithm,

$$5^{\frac{a}{b}} = 2, \quad \text{and so} \quad (5^{\frac{a}{b}})^b = 2^b \quad \text{or, equivalently,} \quad 5^a = 2^b$$

Let
$$N = 5^a = 2^b.$$

Since $b \ge 0$, $N > 2^0 = 1$. Thus we may consider the prime factorization of N.

Because $N = 5^a$, the prime factors of N are all 5. On the other hand, because $N = 2^b$, the prime factors of N are all 2.

This contradicts the unique factorization of integers theorem, which states that the prime factors of any integer greater than 1 are unique except for the order in which they are written.

Hence the supposition is false, and so $\log_5(2)$ is irrational.

27. *a.* All of the following are prime numbers: $N_1 = 2+1 = 3$, $N_2 = 2 \cdot 3+1 = 7$, $N_3 = 2 \cdot 3 \cdot 5+1 = 31$, $N_4 = 2 \cdot 3 \cdot 5 \cdot 7 + 1 = 211$, $N_5 = 2 \cdot 3 \cdot 5 \cdot 7 \cdot 11 + 1 = 2311$. However,

$$N_6 = 2 \cdot 3 \cdot 5 \cdot 7 \cdot 11 \cdot 13 + 1 = 30031 = 59 \cdot 509.$$

Thus the smallest non-prime integer of the given form is 30,031.

b. Each of N_1, N_2, N_3, N_4, and N_5 is prime, and so each is its own smallest prime divisor. Thus $q_1 = N_1$, $q_2 = N_2$, $q_3 = N_3$, $q_4 = N_4$, and $q_5 = N_5$. However, N_6 is not prime and $N_6 = 30031 = 59 \cdot 509$. Since 59 and 509 are primes, the smallest prime divisor of N_6 is $q_6 = 59$.

30. Proof: Let $p_1, p_2, \ldots, p_n$ be distinct prime numbers with $p_1 = 2$ and $n > 1$. *[We must show that $p_1 p_2 \cdots p_n + 1 = 4k + 3$ for some integer k.]* Let

$$N = p_1 p_2 \cdots p_n + 1.$$

By the quotient-remainder theorem, there is an integer k such that

$$N = 4k, \ 4k + 1, \ 4k + 2, \ \text{or} \ 4k + 3.$$

But N is odd (because $p_1 = 2$); hence

$$N = 4k + 1 \quad \text{or} \quad 4k + 3$$

Suppose $N = 4k + 1$. *[We will show that this supposition leads to a contradiction.]*

By substitution,

$$4k + 1 = p_1 p_2 \cdots p_n + 1 \quad \text{and so} \quad 4k = p_1 p_2 \cdots p_n.$$

Thus

$$4 \mid p_1 p_2 \cdots p_n.$$

But $p_1 = 2$ and all of p_2, p_3, $\ldots$, p_n are odd (being prime numbers that are greater than 2). Consequently, there is only one factor of 2 in the prime factorization of $p_1 p_2 \cdots p_n$, and so

$$4 \nmid p_1 p_2 \cdots p_n,$$

which results in a contradiction. Therefore the supposition that $N = 4k + 1$ for some integer k is false, and so *[by elimination]* $N = 4k + 3$ for some integer k *[as was to be shown]*.

33. Existence Proof: When $n = 2$, then

$$n^2 + 2n - 3 = 2^2 + 2 \cdot 2 - 3 = 5,$$

which is prime. Thus there is a prime number of the form $n^2 + 2n - 3$, where n is a positive integer.

Uniqueness Proof (by contradiction): By the existence proof above, we know that when $n = 2$, then $n^2 + 2n - 3$ is prime. Suppose there is another positive integer m, not equal to 2, such that $m^2 + 2m - 3$ is prime. *[We will show that this supposition leads logically to a contradiction.]* By factoring, we see that

$$m^2 + 2m - 3 = (m + 3)(m - 1).$$

Now $m \neq 1$ because otherwise $m^2 + 2m - 3 = 0$, which is not prime. Also $m \neq 2$ by supposition. Thus $m > 2$. Consequently, $m + 3 > 5$ and $m - 1 > 1$, and so $m^2 + 2m - 3$ can be written as a product of two positive integers neither of which is 1 (namely $m + 3$ and $m - 1$). This contradicts the supposition that $m^2 + 2m - 3$ is prime. Hence the supposition is false: there is no integer m other than 2 such that $m^2 + 2m - 3$ is prime.

Uniqueness Proof (direct): Suppose m is any positive integer such that $m^2 + 2m - 3$ is prime. *[We will show that $m = 2$.]* By factoring,

$$m^2 + 2m - 3 = (m + 3)(m - 1).$$

Since $m^2 + 2m - 3$ is prime, either $m + 3 = 1$ or $m - 1 = 1$. Now $m + 3 \neq 1$ because m is positive and if $m + 3 = 1$ then $m = -2$. Thus $m - 1 = 1$, which implies that $m = 2$ *[as was to be shown.]*.

Section 4.8

3. *b.* $z = 6$

12. *Solution 1*: $\gcd(48, 54) = \gcd(6 \cdot 8, 6 \cdot 9) = 6$

    Solution 2: $\gcd(48, 54) = \gcd(2^4 \cdot 3, 2 \cdot 3^3) = 2 \cdot 3 = 6$

15.
$$
\begin{array}{r}
13 \\
832\overline{\smash{\big)}10933} \\
\underline{10816} \\
117
\end{array}
$$
So $10933 = 832 \cdot 13 + 117$, and hence $\gcd(10933, 832) = \gcd(832, 117)$

$$
\begin{array}{r}
7 \\
117\overline{\smash{\big)}832} \\
\underline{819} \\
13
\end{array}
$$
So $832 = 117 \cdot 7 + 13$, and hence $\gcd(832, 117) = \gcd(117, 13)$

$$
\begin{array}{r}
9 \\
13\overline{\smash{\big)}117} \\
\underline{117} \\
0
\end{array}
$$
So $117 = 13 \cdot 9 + 0$, and hence $\gcd(117, 13) = \gcd(13, 0)$

But $\gcd(13, 0) = 13$. So $\gcd(10933, 832) = 13$.

18.

A	5859						
B	1232						
r	1232	931	301	28	21	7	0
a	5859	1232	931	301	28	21	7
b	1232	931	301	28	21	7	0
gcd							7

21. <u>Proof:</u>

Suppose a and b are any integers with $b \neq 0$, and suppose q and r are any integers such that

$$a = bq + r.$$

We must show that

$$\gcd(b, r) \leq \gcd(a, b).$$

Step 1 (proof that any common divisor of b and r is also a common divisor of a and b):

Let c be a common divisor of b and r. Then $c \mid b$ and $c \mid r$, and so by definition of divisibility, there are integers n and m so that

$$b = nc \quad \text{and} \quad r = mc$$

Substitute these values into the equation $a = bq + r$ to obtain

$$a = (nc)q + mc = c(nq + m).$$

But $nq + m$ is an integer, and so by definition of divisibility $c \mid a$. Because we already know that $c \mid b$, we can conclude that c is a common divisor of a and b.

Step 2 (proof that $\gcd(b, r) \leq \gcd(a, b)$):

By step 1, every common divisor of b and r is a common divisor of a and b. It follows that the greatest common divisor of b and r is a common divisor of a and b. But then $\gcd(b, r)$ (being one of the common divisors of a and b) is less than or equal to the greatest common divisor of a and b:

$$\gcd(b, r) \leq \gcd(a, b)$$

[as was to be shown].

24. *a.* <u>Proof</u>: Suppose a and b are integers and $a \geq b > 0$.

Part 1 (proof that every common divisor of a and b is a common divisor of b and $a - b$):

Suppose

$$d \mid a \quad \text{and} \quad d \mid b.$$

Then, by exercise 16 of Section 4.3,

$$d \mid (a - b).$$

Hence d is a common divisor of a and $a - b$.

Part 2 (proof that every common divisor of b and $a - b$ is a common divisor of a and b):

Suppose

$$d \mid b \quad \text{and} \quad d \mid (a - b).$$

Then, by exercise 15 of Section 4.3,

$$a \mid [b + (a - b)].$$

But $b + (a - b) = a$, and so

$$d \mid a.$$

Hence d is a common divisor of a and b.

Part 3 (end of proof):

Because every common divisor of a and b is a common divisor of b and $a - b$, the greatest common divisor of a and b is a common divisor of b and $a - b$ and so is less than or equal to the greatest common divisor of a and $a - b$. Thus

$$\gcd(a, b) \leq \gcd(b, a - b).$$

By similar reasoning,

$$\gcd(b, a - b) \leq \gcd(a, b) \quad \text{and, therefore,} \quad \gcd(a, b) = \gcd(b, a - b).$$

c.

A	768											
B	348											
a	768	420	72					12				0
b	348			276	204	132	60		48	36	24	12
gcd												12

27. <u>Proof</u>: Let a and b be any positive integers.

   Part 1 (proof that if $\text{lcm}(a, b) = b$ ***then*** $a \mid b$***)***: Suppose that

$$\text{lcm}(a, b) = b.$$

By definition of least common multiple,

$$a \mid \text{lcm}(a, b),$$

and so by substitution, $a \mid b$.

   Part 2 (proof that if $a \mid b$ ***then*** $\text{lcm}(a, b) = b$***)***: Suppose that

$$a \mid b.$$

Then since it is also the case that

$$b \mid b,$$

b is a common multiple of a and b. Moreover, because b divides any common multiple of both a and b,

$$\text{lcm}(a, b) = b.$$

Review Guide: Chapter 4

Definitions

- Why is the phrase "if, and only if" used in a definition? *(p. 147)*
- How are the following terms defined?
  - even integer *(p. 147)*
  - odd integer *(p. 147)*
  - prime number *(p. 148)*
  - composite number *(p. 148)*
  - rational number *(p. 163)*
  - divisibility of one integer by another *(p. 170)*
  - n div d and n mod d *(p. 181)*
  - the floor of a real number *(p. 191)*
  - the ceiling of a real number *(p. 191)*
  - greatest common divisor of two integers *(p. 220)*

Proving an Existential Statement/Disproving a Universal Statement

- How do you determine the truth of an existential statement? *(p. 148)*
- What does it mean to "disprove" a statement? *(p. 149)*
- What is disproof by counterexample? *(p. 149)*
- How do you establish the falsity of a universal statement? *(p. 149)*

Proving a Universal Statement/Disproving an Existential Statement

- If a universal statement is defined over a small, finite domain, how do you use the method of exhaustion to prove that it is true? *(p. 150)*
- What is the method of generalizing from the generic particular? *(p. 151)*
- If you use the method of direct proof to prove a statement of the form "$\forall x$, if $P(x)$ then $Q(x)$", what do you suppose and what do you have to show? *(p. 152)*
- What are the guidelines for writing proofs of universal statements? *(pp. 155-156)*
- What are some common mistakes people make when writing mathematical proofs? *(pp. 157-158)*
- How do you disprove an existential statement? *(p. 159)*
- What is the method of proof by division into cases? *(p. 184)*
- What is the triangle inequality? *(p. 188)*
- If you use the method of proof by contradiction to prove a statement, what do you suppose and what do you have to show? *(p. 198)*
- If you use the method of proof by contraposition to prove a statement of the form "$\forall x$, if $P(x)$ then $Q(x)$", what do you suppose and what do you have to show? *(p. 202)*
- Are you able to use the various methods of proof and disproof to establish the truth or falsity of statements about odd and even integers *(pp. 154,199)*, prime numbers *(pp. 159,210)*, rational and irrational numbers *(pp. 165,166,201,208,209)*, divisibility of integers *(pp. 171,173-175,184,186,202,203)*, absolute value *(pp. 187-188)*, and the floor and ceiling of a real number *(pp. 194-196)*?

Some Important Theorems and Algorithms

- What is the transitivity of divisibility theorem? *(p. 173)*
- What is the theorem about divisibility by a prime number? *(p. 174)*

- What is the unique factorization of integers theorem? (This theorem is also called the fundamental theorem of arithmetic.) *(p. 176)*
- What is the quotient-remainder theorem? Can you apply it to specific situations? *(p. 180)*
- What is the theorem about the irrationality of the square root of 2? Can you prove this theorem? *(p. 208)*
- What is the theorem about the infinitude of the prime numbers? Can you prove this theorem? *(p. 210)*
- What is the division algorithm ? *(p. 219)*
- What is the Euclidean algorithm? *(pp. 220,224)*
- How do you use the Euclidean algorithm to compute the greatest common divisor of two positive integers? *(p. 223)*

Notation for Algorithms

- How is an assignment statement executed? *(p. 214)*
- How is an **if-then** statement executed? *(p. 215)*
- How is an **if-then-else** statement executed? *(p. 215)*
- How are the statements **do** and **end do** used in an algorithm? *(p. 215)*
- How is a **while** loop executed? *(p. 216)*
- How is a **for-next** loop executed? *(p. 217)*
- How do you construct a trace table for a segment of an algorithm? *(pp. 217,219)*

Chapter 5: Sequences, Mathematical Induction, and Recursion

The first section of this chapter introduces the notation for sequences, summations, products, and factorial. The section is intended to help you learn to recognize patterns so as to be able, for instance, to transform expanded versions of sums into summation notation, and to handle subscripts, particularly to change variables for summations and to distinguish index variables from variables that are constant with respect to a summation.

The second, third, and fourth sections of the chapter treat mathematical induction. The ordinary form is discussed in Sections 5.2 and 5.3 and the strong form in Section 5.4. Because of the importance of mathematical induction in discrete mathematics, a wide variety of examples is given to help you become comfortable with using the technique in many different situations. Section 5.5 then shows how to use a variation of mathematical induction to prove the correctness of an algorithm. Sections 5.6-5.8 deal with recursively defined sequences, both how to analyze a situation using recursive thinking to obtain a sequence that describes the situation and how to find an explicit formula for the sequence once it has been defined recursively. Section 5.9 applies recursive thinking to the question of defining a set, and it describes the technique of structural induction, which is the variation of mathematical induction that can be used to verify properties of a set that has been defined. Section 5.9 also introduces the concept of a recursively defined function.

The logic of ordinary mathematical induction can be described by relating it to the logic discussed in Chapters 2 and 3. The main point is that the inductive step establishes the truth of a sequence of if-then statements. Together with the basis step, this sequence gives rise to a chain of inferences that lead to the desired conclusion. More formally:

Suppose
1. $P(1)$ is true; and
2. for all integers $k \geq 1$, if $P(k)$ is true then $P(k + 1)$ is true.

The truth of statement (2) implies, according to the law of universal instantiation, that no matter what particular integer $k \geq 1$ is substituted in place of k, the statement "If $P(k)$ then $P(k+1)$" is true. The following argument, therefore, has true premises, and so by modus ponens it has a true conclusion:

$$
\begin{array}{lll}
& \text{If } P(1) \text{ then } P(2). & \text{by 2 and universal instantiation} \\
& P(1) & \text{by 1} \\
\therefore & P(2) & \text{by modus ponens}
\end{array}
$$

Similar reasoning gives the following chain of arguments, each of which has a true conclusion by modus ponens:

$$
\begin{array}{ll}
& \text{If } P(2) \text{ then } P(3). \\
& P(2) \\
\therefore & P(3) \\
& \text{If } P(3) \text{ then } P(4). \\
& P(3) \\
\therefore & P(4) \\
& \text{If } P(4) \text{ then } P(5). \\
& P(4) \\
\therefore & P(5) \\
& \text{And so forth.}
\end{array}
$$

Thus no matter how large a positive integer n is specified, the truth of $P(n)$ can be deduced as the final conclusion of a (possibly very long) chain of arguments continuing those shown above.

48

Section 5.1

6. $f_1 = \left\lfloor \dfrac{1}{4} \right\rfloor \cdot 4 = 0 \cdot 4 = 0, \quad f_2 = \left\lfloor \dfrac{2}{4} \right\rfloor \cdot 4 = 0 \cdot 4 = 0, \quad f_3 = \left\lfloor \dfrac{3}{4} \right\rfloor \cdot 4 = 0 \cdot 4 = 4,$

$f_4 = \left\lfloor \dfrac{4}{4} \right\rfloor \cdot 4 = 1 \cdot 4 = 4$

9.
$$\begin{aligned}
h_1 &= 1 \cdot \lfloor \log_2 1 \rfloor &&= 1 \cdot 0 \\
h_2 &= 2 \cdot \lfloor \log_2 2 \rfloor &&= 2 \cdot 1 \\
h_3 &= 3 \cdot \lfloor \log_2 3 \rfloor &&= 3 \cdot 1 \\
h_4 &= 4 \cdot \lfloor \log_2 4 \rfloor &&= 4 \cdot 2 \\
h_5 &= 5 \cdot \lfloor \log_2 5 \rfloor &&= 5 \cdot 2 \\
h_6 &= 6 \cdot \lfloor \log_2 6 \rfloor &&= 6 \cdot 2 \\
h_7 &= 7 \cdot \lfloor \log_2 7 \rfloor &&= 7 \cdot 2 \\
h_8 &= 8 \cdot \lfloor \log_2 8 \rfloor &&= 8 \cdot 3 \\
h_9 &= 9 \cdot \lfloor \log_2 9 \rfloor &&= 9 \cdot 3 \\
h_{10} &= 10 \cdot \lfloor \log_2 10 \rfloor &&= 10 \cdot 3 \\
h_{11} &= 11 \cdot \lfloor \log_2 11 \rfloor &&= 11 \cdot 3 \\
h_{12} &= 12 \cdot \lfloor \log_2 12 \rfloor &&= 12 \cdot 3 \\
h_{13} &= 13 \cdot \lfloor \log_2 13 \rfloor &&= 13 \cdot 3 \\
h_{14} &= 14 \cdot \lfloor \log_2 14 \rfloor &&= 14 \cdot 3 \\
h_{15} &= 15 \cdot \lfloor \log_2 15 \rfloor &&= 15 \cdot 3
\end{aligned}$$

When n is an integral power of 2, h_n is n times the exponent of that power. For instance, $8 = 2^3$ and $h_8 = 8 \cdot 3$. If m and n are integers and $2^m \le n < 2^{m+1}$, then $h_n = n \cdot m$.

15. $a_n = (-1)^{n-1} \left(\dfrac{n-1}{n} \right)$ for all integers $n \ge 1$ (There are other correct answers for this exercise.)

18. $e.$ $\prod_{k=2}^{2} a_k = a_2 = -2$

21. $\displaystyle\sum_{m=0}^{3} \dfrac{1}{2^m} = \dfrac{1}{2^0} + \dfrac{1}{2^1} + \dfrac{1}{2^2} + \dfrac{1}{2^3} = 1 + \dfrac{1}{2} + \dfrac{1}{4} + \dfrac{1}{8} = \dfrac{15}{8}$

24. $\displaystyle\sum_{j=0}^{0} (j+1) \cdot 2^j = (0+1) \cdot 2^0 = 1 \cdot 1 = 1$

30. $\displaystyle\sum_{j=1}^{n} j(j+1) = 1 \cdot 2 + 2 \cdot 3 + 3 \cdot 4 + \cdots + n \cdot (n+1)$ (There are other correct answers for this exercise.)

36. $\left(\dfrac{1 \cdot 2}{3 \cdot 4} \right) = \dfrac{1}{6}$

39. $\displaystyle\sum_{m=1}^{n+1} m(m+1) = \sum_{m=1}^{n} m(m+1) + (n+1)((n+1)+1)$

42. $\displaystyle\sum_{m=0}^{n} (m+1)2^m + (n+2)2^{n+1} = \sum_{m=0}^{n+1} (m+1)2^m$

Exercises 45 and 48 have more than one correct answer.

45. $\displaystyle\prod_{i=2}^{4} (i^2 - 1)$

48. $\displaystyle\prod_{j=1}^{4}(1-t^{j})$

54. When $k=1$, $i=1+1=2$. When $k=n$, $i=n+1$. Since $i=k+1$, then $k=i-1$. So

$$\frac{k}{k^2+4}=\frac{(i-1)}{(i-1)^2+4}=\frac{i-1}{i^2-2i+1+4}=\frac{i-1}{i^2-2i+5}.$$

Therefore,

$$\prod_{k=1}^{n}(\frac{k}{k^2+4})=\prod_{i=2}^{n+1}(\frac{i-1}{i^2-2i+5}).$$

57. When $i=1$, $j=1-1=0$. When $i=n-1$, $j=n-2$. Since $j=i-1$, then $i=j+1$. So

$$\frac{i}{(n-i)^2}=\frac{j+1}{(n-(j+1))^2}=\frac{j+1}{(n-j-1)^2}.$$

Therefore,

$$\sum_{i=1}^{n-1}(\frac{i}{(n-i)^2})=\sum_{j=0}^{n-2}(\frac{j+1}{(n-j-1)^2}).$$

60. By Theorem 5.1.1,

$$2\cdot\sum_{k=1}^{n}(3k^2+4)+5\cdot\sum_{k=1}^{n}(2k^2-1)\;\;=\;\;\sum_{k=1}^{n}2(3k^2+4)+\sum_{k=1}^{n}5(2k^2-1)$$

$$=\;\;\sum_{k=1}^{n}(6k^2+8)+\sum_{k=1}^{n}(10k^2-5)$$

$$=\;\;\sum_{k=1}^{n}(6k^2+8+10k^2-5)$$

$$=\;\;\sum_{k=1}^{n}(16k^2+3).$$

63. $\displaystyle\frac{6!}{8!}=\frac{6!}{8\cdot7\cdot6!}=\frac{1}{56}$

72. $\displaystyle\binom{7}{4}=\frac{7!}{4!(7-4)!}=\frac{7\cdot6\cdot5\cdot4\cdot3\cdot2\cdot1}{(4\cdot3\cdot2\cdot1)(3\cdot2\cdot1)}=\frac{7\cdot6\cdot5}{(3\cdot2\cdot1)}=35$

87. Let a nonnegative integer a be given. Divide a by 16 using the quotient-remainder theorem to obtain a quotient $q[0]$ and a remainder $r[0]$. If the quotient is nonzero, divide by 16 again to obtain a quotient $q[1]$ and a remainder $r[1]$. Continue this process until a quotient of 0 is obtained. The remainders calculated in this way are the hexadecimal digits of a:

$$a_{10}=(r[k]r[k-1]\ldots r[2]r[1]r[0])_{16}.$$

90.

```
          0      R. 8 = 8₁₆
    16│   8      R. 15  = F₁₆
    16│  143     R. 13 = D₁₆
    16│ 2301                Hence 2301₁₀ = 8FD₁₆.
```

R. 8 $=8_{16}$
R. 15 $=F_{16}$
R. 13 $=D_{16}$
Hence $2301_{10}=8FD_{16}$.

Section 5.2

9. <u>Proof (by mathematical induction)</u>: Let the property $P(n)$ be the equation

$$4^3 + 4^4 + 4^5 + \cdots + 4^n = \frac{4(4^n - 16)}{3}. \qquad \leftarrow P n)$$

Show that $P(3)$ is true: $P(3)$ is true because the left-hand side is $4^3 = 64$ and the right-hand side is $\dfrac{4(4^3 - 16)}{3} = \dfrac{4(64 - 16)}{3} = \dfrac{4 \cdot 48}{3} = 64$ also.

Show that for all integers $k \geq 3$, if $P(k)$ is true then $P(k+1)$ is true: Let k be any integer with $k \geq 3$, and suppose that

$$4^3 + 4^4 + 4^5 + \cdots + 4^k = \frac{4(4^k - 16)}{3}. \qquad \leftarrow \begin{array}{l} P(k) \\ \text{inductive hypothesis} \end{array}$$

We must show that

$$4^3 + 4^4 + 4^5 + \cdots + 4^{k+1} = \frac{4(4^{k+1} - 16)}{3}. \qquad \leftarrow P(k+1)$$

Now the left-hand side of $P(k+1)$ is

$$
\begin{aligned}
4^3 + 4^4 + 4^5 + \cdots + 4^{k+1} \;&=\; 4^3 + 4^4 + 4^5 + \cdots + 4^k + 4^{k+1} \\
&\qquad \text{by making the next-to-last term explicit} \\
&=\; \frac{4(4^k - 16)}{3} + 4^{k+1} \\
&\qquad \text{by inductive hypothesis} \\
&=\; \frac{4^{k+1} - 64}{3} + \frac{3 \cdot 4^{k+1}}{3} \\
&\qquad \text{by creating a common denominator} \\
&=\; \frac{4 \cdot 4^{k+1} - 64}{3} \\
&\qquad \text{by adding the fractions} \\
&=\; \frac{4(4^{k+1} - 16)}{3} \\
&\qquad \text{by factoring out the 4,}
\end{aligned}
$$

and this is the right-hand side of $P(k+1)$ *[as was to be shown]*.

[Since both the basis and the inductive steps have been proved, we conclude that $P(n)$ is true for all integers $n \geq 3$.]

12. <u>Proof (by mathematical induction)</u>: Let the property $P(n)$ be the equation

$$\frac{1}{1 \cdot 2} + \frac{1}{2 \cdot 3} + \cdots + \frac{1}{n(n+1)} = \frac{n}{n+1}. \qquad \leftarrow P n)$$

Show that $P(1)$ is true: $P(1)$ is true because the left-hand side equals $\dfrac{1}{1 \cdot 2} = \dfrac{1}{2}$ and the right-hand side equals $\dfrac{1}{1+1} = \dfrac{1}{2}$ also.

Show that for all integers $k \geq 1$, if $P(k)$ is true then $P(k+1)$ is true: Let k be any integer with $k \geq 1$ and suppose that

$$\frac{1}{1 \cdot 2} + \frac{1}{2 \cdot 3} + \cdots + \frac{1}{k(k+1)} = \frac{k}{k+1}. \qquad \leftarrow \begin{array}{l} P(k) \\ \text{inductive hypothesis} \end{array}$$

We must show that

$$\frac{1}{1\cdot 2} + \frac{1}{2\cdot 3} + \cdots + \frac{1}{(k+1)((k+1)+1)} = \frac{k+1}{(k+1)+1},$$

or, equivalently,

$$\frac{1}{1\cdot 2} + \frac{1}{2\cdot 3} + \cdots + \frac{1}{(k+1)(k+2)} = \frac{k+1}{k+2}. \quad \leftarrow P(k+1)$$

The left-hand side of $P(k+1)$ is

$$\frac{1}{1\cdot 2} + \frac{1}{2\cdot 3} + \cdots + \frac{1}{(k+1)(k+2)}$$

$$= \frac{1}{1\cdot 2} + \frac{1}{2\cdot 3} + \cdots + \frac{1}{k(k+1)} + \frac{1}{(k+1)(k+2)}$$
by making the next-to-last term explicit

$$= \frac{k}{k+1} + \frac{1}{(k+1)(k+2)}$$
by inductive hypothesis

$$= \frac{k(k+2)}{(k+1)(k+2)} + \frac{1}{(k+1)(k+2)}$$
by creating a common denominator

$$= \frac{k^2 + 2k + 1}{(k+1)(k+2)}$$
by adding the fractions

$$= \frac{(k+1)^2}{(k+1)(k+2)}$$
because $k^2 + 2k + 1 = (k+1)^2$

$$= \frac{k+1}{k+2}$$
by canceling $(k+1)$ from numerator and denominator,

and this is the right-hand side of $P(k+1)$ *[as was to be shown]*

15. Proof (by mathematical induction): Let the property $P(n)$ be the equation

$$\sum_{i=1}^{n} i(i!) = (n+1)! - 1. \quad \leftarrow P(n)$$

Show that $P(1)$ is true: We must show that $\sum_{i=1}^{1} i(i!) = (1+1)! - 1$. But the left-hand side of this equation is $\sum_{i=1}^{1} i(i!) = 1\cdot(1!) = 1$ and the right-hand side is $(1+1)! - 1 = 2! - 1 = 2 - 1 = 1$ also. So $P(1)$ is true.

Show that for all integers $k \geq 1$, if $P(k)$ is true then $P(k+1)$ is true: Let k be any integer with $k \geq 1$, and suppose that

$$\sum_{i=1}^{k} i(i!) = (k+1)! - 1. \quad \leftarrow \begin{array}{l} P(k) \\ \text{inductive hypothesis} \end{array}$$

We must show that

$$\sum_{i=1}^{k+1} i(i!) = ((k+1)+1)! - 1,$$

or, equivalently,

$$\sum_{i=1}^{k+1} i(i!) = (k+2)! - 1. \quad \leftarrow P(k+1)$$

The left-hand side of $P(k+1)$ is

$$\sum_{i=1}^{k+1} i(i!) \quad = \quad \sum_{i=1}^{k} i(i!) + (k+1)((k+1)!) \qquad \text{by writing the } (k+1)\text{st term separately}$$

$$= \quad [(k+1)! - 1] + (k+1)((k+1)!) \quad \text{by inductive hypothesis}$$

$$= \quad ((k+1)!)(1 + (k+1)) - 1 \qquad \text{by combining the terms with}$$
$$\qquad\qquad\qquad\qquad\qquad\qquad\qquad\qquad \text{the common factor } (k+1)!$$

$$= \quad (k+1)!(k+2) - 1$$

$$= \quad (k+2)! - 1 \qquad\qquad\qquad\qquad \text{by algebra,}$$

and this is the right-hand side of $P(k+1)$ *[as was to be shown]*.

18. <u>Proof (by mathematical induction)</u>: Let the property $P(n)$ be the equation

$$\sin x + \sin 3x + \cdots + \sin(2n-1)x = \frac{1 - \cos 2nx}{2 \sin x}. \qquad \leftarrow Pn)$$

Show that $P(1)$ ***is true***: $P(1)$ is true because the left-hand side equals $\sin x$, and the right-hand side equals

$$\frac{1 - \cos 2x}{2 \sin x} = \frac{1 - \cos^2 + \sin^2 x}{2 \sin x} = \frac{2 \sin^2 x}{2 \sin x} = \sin x.$$

Show that for all integers $k \geq 1$, ***if*** $P(k)$ ***is true then*** $P(k+1)$ ***is true***: Let k be any integer with $k \geq 1$ and suppose that

$$\sin x + \sin 3x + \cdots + \sin(2k-1)x = \frac{1 - \cos 2kx}{2 \sin x}. \qquad \leftarrow \begin{array}{l} P(k) \\ \text{inductive hypothesis} \end{array}$$

We must show that
$$\sin x + \sin 3x + \cdots + \sin(2(k+1) - 1)x = \frac{1 - \cos 2(k+1)x}{2 \sin x},$$

or, equivalently,
$$\sin x + \sin 3x + \cdots + \sin(2k+1)x = \frac{1 - \cos 2(k+1)x}{2 \sin x}. \qquad \leftarrow P(k+1)$$

When the next-to-last term of the the left-hand side of $P(k+1)$ is made explicit, the left-hand side becomes
$$\sin x + \sin 3x + \cdots + \sin(2k-1)x + \sin(2k+1)x$$

$$= \quad \frac{1 - \cos 2kx}{2 \sin x} + \sin(2k+1)x \qquad\qquad \text{by inductive hypothesis}$$

$$= \quad \frac{1 - \cos 2kx}{2 \sin x} + \frac{2 \sin x \sin(2kx + x)}{2 \sin x} \qquad \text{by creating a common denominator}$$

$$= \quad \frac{1 - \cos 2kx + 2 \sin x \sin(2kx + x)}{2 \sin x} \qquad \text{by adding fractions}$$

$$= \quad \frac{1 - \cos 2kx + 2 \sin x[\sin(2kx)\cos x + \cos(2kx)\sin x]}{2 \sin x} \qquad \text{by the addition formula for sine}$$

$$= \quad \frac{1 - \cos 2kx + 2 \sin x \sin(2kx)\cos x + 2 \sin^2 x \cos(2kx)}{2 \sin x} \qquad \text{by multiplying out}$$

$$= \quad \frac{1 + \cos 2kx(2 \sin^2 x - 1) + 2 \sin x \cos x \sin(2kx)}{2 \sin x} \qquad \text{by combining like terms}$$

$$= \quad \frac{1 + \cos 2kx(-\cos 2x) + \sin 2x \sin(2kx)}{2 \sin x} \qquad \text{by the formulas for } \cos 2x \text{ and } \sin 2x$$

$$= \quad \frac{1 - (\cos 2kx \cos 2x - \sin 2x \sin(2kx))}{2 \sin x} \qquad \text{by factoring out } -1$$

$$= \quad \frac{1 - \cos(2kx + 2x)}{2 \sin x} \qquad \text{by the addition formula for cosine}$$

$$= \quad \frac{1 - \cos(2(k+1)x)}{2 \sin x} \qquad \text{by factoring out } 2x,$$

and this is the right-hand side of $P(k+1)$ *[as was to be shown]*.

21. $5 + 10 + 15 + 20 + \cdots + 300 = 5(1 + 2 + 3 + \cdots + 60) = 5\left(\dfrac{60 \cdot 61}{2}\right) = 9150.$

27. *Solution 1:*

$$
\begin{aligned}
5^3 + 5^4 + 5^5 + \cdots + 5^k &= 5^3(1 + 5 + 5^2 + \cdots + 5^{k-3}) \\
&= 5^3\left(\frac{5^{(k-3)+1} - 1}{5 - 1}\right) = \frac{5^3(5^{k-2} - 1)}{4}
\end{aligned}
$$

Solution 2:

$$
\begin{aligned}
5^3 + 5^4 + 5^5 + \cdots + 5^k &= 1 + 5 + 5^2 + \cdots + 5^k - (1 + 5 + 5^2) \\
&= \frac{5^{k+1} - 1}{5 - 1} - 31 = \frac{5^{k+1} - 1}{4} - 31
\end{aligned}
$$

Note that the expression obtained in solution 2 can be transformed into the one obtained in solution 1:

$$
\frac{5^{k+1} - 1}{4} - 31 = \frac{5^{k+1} - 1}{4} - \frac{31 \cdot 4}{4} = \frac{5^{k+1} - 125}{4} = \frac{5^{k+1} - 5^3}{4} = \frac{5^3(5^{k-2} - 1)}{4}.
$$

30. $(a + md) + (a + (m+1)d) + (a + (m+2)d) + \cdots + (a + (m+n)d)$

$$
\begin{aligned}
&= (a + md) + (a + md + d) + (a + md + 2d) + \cdots + (a + md + nd) \\
&= \underbrace{((a + md) + (a + md) + \cdots + (a + md))}_{n + 1 \text{ terms}} + d(1 + 2 + 3 + \cdots + n) \\
&= (n+1)(a + md) + d\left(\frac{n(n+1)}{2}\right) \qquad \text{by Theorem 5.2.2} \\
&= (a + md + \frac{n}{2}d)(n+1) \\
&= [a + (m + \frac{n}{2})d](n+1)
\end{aligned}
$$

Any one of the last three equations or their algebraic equivalents could be considered a correct answer.

36. <u>Proof:</u> Suppose m and n are any positive integers such that m is odd. By definition of odd, $m = 2q + 1$ for some integer k, and so, by Theorems 5.1.1 and 5.2.2,

$$
\sum_{k=0}^{m-1}(n+k) = \sum_{k=0}^{(2q+1)-1}(n+k) = \sum_{k=0}^{2q}(n+k) = \sum_{k=0}^{2q}n + \sum_{k=0}^{2q}k = (2q+1)n + \sum_{k=1}^{2q}k
$$

$$
= (2q+1)n + \frac{2q(2q+1)}{2} = (2q+1)n + q(2q+1) = (2q+1)(n+q) = m(n+q).
$$

But $n + q$ is an integer because it is a sum of integers. Hence, by definition of divisibility, $\sum_{k=0}^{m-1}(n+k)$ is divisible by m.

Note: If m is even, the property is no longer true. For example, if $n = 1$ and $m = 2$, then $\sum_{k=0}^{m-1}(n+k) = \sum_{k=0}^{2-1}(1+k) = 1 + 2 = 3$, and 3 is not divisible by 2.

Section 5.3

9. <u>Proof (by mathematical induction):</u> Let the property $P(n)$ be the sentence

$$7^n - 1 \text{ is divisible by } 6.$$

We will prove that $P(n)$ is true for all integers $n \geq 0$.

Show that P(0) is true: $P(0)$ is true because because $7^0 - 1 = 1 - 1 = 0$ and 0 is divisible by 6 (since $0 = 0 \cdot 6$).

Show that for all integers $k \geq 0$, if $P(k)$ is true then $P(k+1)$ is true: Let k be any integer with $k \geq 0$, and suppose

$$7^k - 1 \text{ is divisible by 6.} \quad \leftarrow \text{ inductive hypothesis}$$

We must show that
$$7^{k+1} - 1 \text{ is divisible by 6.}$$

By definition of divisibility, the inductive hypothesis is equivalent to the statement

$$7^k - 1 = 6r$$

for some integer r. Then

$$
\begin{aligned}
7^{k+1} - 1 &= 7 \cdot 7^k - 1 \\
&= (6+1)7^k - 1 \\
&= 6 \cdot 7^k + (7^k - 1) \quad \text{by algebra} \\
&= 6 \cdot 7^k + 6r \quad \text{by inductive hypothesis} \\
&= 6(7^k + r) \quad \text{by algebra.}
\end{aligned}
$$

Now $7^k + r$ is an integer because products and sums of integers are integers. Thus, by definition of divisibility, $7^{k+1} - 1$ is divisible by 6 *[as was to be shown]*.

12. <u>Proof (by mathematical induction)</u>: Let the property $P(n)$ be the sentence

$$7^n - 2^n \text{ is divisible by 5.}$$

We will prove that $P(n)$ is true for all integers $n \geq 0$.

Show that P(0) is true: $P(0)$ is true because $7^0 - 2^0 = 0 - 0 = 0$ and 0 is divisible by 5 (since $0 = 5 \cdot 0$).

Show that for all integers $k \geq 0$, if $P(k)$ is true then $P(k+1)$ is true: Let k be any integer with $k \geq 0$, and suppose

$$7^k - 2^k \text{ is divisible by 5.} \quad \leftarrow \text{ inductive hypothesis}$$

We must show that
$$7^{k+1} - 2^{k+1} \text{ is divisible by 5.}$$

By definition of divisibility, the inductive hypothesis is equivalent to the statement $7^k - 2^k = 5r$ for some integer r. Then

$$
\begin{aligned}
7^{k+1} - 2^{k+1} &= 7 \cdot 7^k - 2 \cdot 2^k \\
&= (5+2) \cdot 7^k - 2 \cdot 2^k \\
&= 5 \cdot 7^k + 2 \cdot 7^k - 2 \cdot 2^k \\
&= 5 \cdot 7^k + 2(7^k - 2^k) \quad \text{by algebra} \\
&= 5 \cdot 7^k + 2 \cdot 5r \quad \text{by inductive hypothesis} \\
&= 5(7^k + 2r) \quad \text{by algebra.}
\end{aligned}
$$

Now $7^k + 2r$ is an integer because products and sums of integers are integers. Therefore, by definition of divisibility, $7^{k+1} - 2^{k+1}$ is divisible by 5 *[as was to be shown]*.

15. <u>Proof (by mathematical induction)</u>: Let the property $P(n)$ be the sentence

$$n(n^2 + 5) \text{ is divisible by 6.}$$

We will prove that $P(n)$ is true for all integers $n \geq 0$.

Show that $P(0)$ is true: $P(0)$ is true because $0(0^2 + 5) = 0$ and 0 is divisible by 6.

Show that for all integers $k \geq 0$, if $P(k)$ is true then $P(k+1)$ is true: Let k be any integer with $k \geq 0$, and suppose

$$k(k^2 + 5) \text{ is divisible by 6.} \qquad \leftarrow \text{ inductive hypothesis}$$

We must show that

$$(k+1)((k+1)^2 + 5) \text{ is divisible by 6.}$$

By definition of divisibility $k(k^2 + 5) = 6r$ for some integer r. Then

$$
\begin{aligned}
(k+1)((k+1)^2 + 5) &= (k+1)(k^2 + 2k + 1 + 5) \\
&= (k+1)(k^2 + 2k + 6) \\
&= k^3 + 2k^2 + 6k + k^2 + 2k + 6 \\
&= k^3 + 3k^2 + 8k + 6 \\
&= (k^3 + 5k) + (3k^2 + 3k + 6) \\
&= k(k^2 + 5) + (3k^2 + 3k + 6) \qquad \text{by algebra} \\
&= 6r + 3(k^2 + k) + 6, \qquad\qquad \text{by inductive hypothesis.}
\end{aligned}
$$

Now $k(k+1)$ is a product of two consecutive integers. By Theorem 4.4.2 one of these is even, and so *[by properties 1 and 4 of Example 4.2.3]* the product $k(k+1)$ is even. Hence $k(k+1) = 2s$ for some integer s. Thus

$$6r + 3(k^2 + k) + 6 = 6r + 3(2s) + 6 = 6(r + s + 1).$$

By substitution, then,

$$(k+1)((k+1)^2 + 5) = 6(r + s + 1),$$

which is divisible by 6 because $r+s+1$ is an integer. Therefore, $(k+1)((k+1)^2 + 5)$ is divisible by 6 *[as was to be shown]*.

18. Proof (by mathematical induction): Let the property $P(n)$ be the inequality $5^n + 9 < 6^n$.

    We will prove that $P(n)$ is true for all integers $n \geq 2$.

    Show that $P(2)$ is true: $P(2)$ is true because the left-hand side is $5^2 + 9 = 25 + 9 = 34$ and the right-hand side is $6^2 = 36$, and $34 < 36$.

    Show that for all integers $k \geq 2$, if $P(k)$ is true then $P(k+1)$ is true: Let k be any integer with $k \geq 2$, and suppose

    $$5^k + 9 < 6^k. \qquad \leftarrow \text{ inductive hypothesis}$$

    We must show that

    $$5^{k+1} + 9 < 6^{k+1}.$$

    Multiplying both sides of the inequality in the inductive hypothesis by 5 gives

    $$5(5^k + 9) < 5 \cdot 6^k. \text{ (*)}$$

    Note that

    $$5^{k+1} + 9 < 5^{k+1} + 45 = 5(5^k + 9) \quad \text{and} \quad 5 \cdot 6^k < 6^{k+1}. \text{ (**)}$$

    Thus, by by the transitive property of order, (*), and (**),

    $$5^{k+1} + 9 < 5(5^k + 9) \quad \text{and} \quad 5(5^k + 9) < 5 \cdot 6^k \quad \text{and} \quad 5 \cdot 6^k < 6^{k+1}.$$

    So, by the transitive property of order,

    $$5^{k+1} + 9 < 6^{k+1}$$

    [as was to be shown].

21. <u>Proof (by mathematical induction)</u>: Let the property $P(n)$ be the inequality

$$\sqrt{n} < \frac{1}{\sqrt{1}} + \frac{1}{\sqrt{2}} + \frac{1}{\sqrt{3}} + \cdots + \frac{1}{\sqrt{n}}.$$

We will prove that $P(n)$ is true for all integers $n \geq 2$.

Show that $P(2)$ is true: To show that $P(2)$ is true we must show that

$$\sqrt{2} < \frac{1}{\sqrt{1}} + \frac{1}{\sqrt{2}}.$$

But this inequality is true if, and only if,

$$2 < \sqrt{2} + 1$$

(by multiplying both sides by $\sqrt{2}$). And this is true if, and only if,

$$1 < \sqrt{2}$$

(by subtracting 1 on both sides). But $1 < \sqrt{2}$, and so $P(2)$ is true.

Show that for all integers $k \geq 2$, if $P(k)$ is true then $P(k+1)$ is true: Let k be any integer with $k \geq 2$, and suppose

$$\sqrt{k} < \frac{1}{\sqrt{1}} + \frac{1}{\sqrt{2}} + \frac{1}{\sqrt{3}} + \cdots + \frac{1}{\sqrt{k}}. \qquad \leftarrow \text{ inductive hypothesis}$$

We must show that

$$\sqrt{k+1} < \frac{1}{\sqrt{1}} + \frac{1}{\sqrt{2}} + \frac{1}{\sqrt{3}} + \cdots + \frac{1}{\sqrt{k+1}}.$$

But for each integer $k \geq 2$,

$$\sqrt{k} < \sqrt{k+1} \ (*),$$

and multiplying both sides of (*) by $\sqrt{k}$ gives

$$k < \sqrt{k} \cdot \sqrt{k+1}.$$

Adding 1 to both sides gives

$$k + 1 < \sqrt{k} \cdot \sqrt{k+1} + 1,$$

and dividing both sides by $\sqrt{k+1}$ gives

$$\sqrt{k+1} < \sqrt{k} + \frac{1}{\sqrt{k+1}}.$$

By substitution from the inductive hypothesis, then,

$$\sqrt{k+1} < \frac{1}{\sqrt{1}} + \frac{1}{\sqrt{2}} + \frac{1}{\sqrt{3}} + \cdots + \frac{1}{\sqrt{k}} + \frac{1}{\sqrt{k+1}}$$

[as was to be shown].

(*) *Note*: Strictly speaking, the reason for this claim is that $k < k+1$ and for all positive real numbers a and b, if $a < b$, then $\sqrt{a} < \sqrt{b}$.

27. <u>Proof (by mathematical induction)</u>: According to the definition of $d_1, d_2, d_3, \ldots$, we have that $d_1 = 2$ and $d_k = \dfrac{d_{k-1}}{k}$ for all integers $k \geq 2$. Let the property $P(n)$ be the equation

$$d_n = \frac{2}{n!}.$$

We will prove that $P(n)$ is true for all integers $n \geq 1$.

Show that $P(1)$ is true: To show that $P(1)$ is true we must show that $d_1 = \frac{2}{1!}$. But $\frac{2}{1!} = 2$ and $d_1 = 2$ (by definition of $d_1, d_2, d_3, \ldots$). So the property holds for $n = 1$.

Show that for all integers $k \geq 1$, if $P(k)$ is true then $P(k+1)$ is true: Let k be any integer with $k \geq 1$, and suppose that

$$d_k = \frac{2}{k!}. \qquad \leftarrow \text{ inductive hypothesis}$$

We must show that

$$d_{k+1} = \frac{2}{(k+1)!}.$$

But the left-hand side of this equation is

$$
\begin{aligned}
d_{k+1} &= \frac{d_k}{k+1} && \text{by definition of } d_1, d_2, d_3, \ldots \\[2mm]
&= \frac{\frac{2}{k!}}{k+1} && \text{by inductive hypothesis} \\[2mm]
&= \frac{2}{(k+1)k!} \\[2mm]
&= \frac{2}{(k+1)!} && \text{by the algebra of fractions,}
\end{aligned}
$$

which is the right-hand side of the equation. *[This is what was to be shown]*.

33.

36. <u>Proof by mathematical induction</u>: Let the property $P(n)$ be the sentence

> In any round-robin tournament involving n teams, it is possible to label the teams $T_1, T_2, T_3, \ldots, T_n$ so that for all $i = 1, 2, 3, \ldots, n-1$, T_i beats T_{i+1}.

We will prove that $P(n)$ is true for all integers $n \geq 2$.

Show that $P(2)$ is true: Consider any round-robin tournament involving two teams. By definition of round-robin tournament, these teams play each other exactly once. Let T_1 be the winner and T_2 the loser of this game. Then T_1 beats T_2, and so the labeling is as required for $P(2)$ to be true.

Show that for all integers $k \geq 2$, if $P(k)$ is true then $P(k+1)$ is true: Let k be any integer with $k \geq 2$ and suppose that

> In any round-robin tournament involving k teams, it is possible to label the $\leftarrow$ inductive teams $T_1, T_2, T_3, \ldots, T_k$ so that for all $i = 1, 2, 3, \ldots, k-1$, T_i beats T_{i+1}.   hypothesis

We must show that

> In any round-robin tournament involving $k + 1$ teams, it is possible to label the teams $T_1, T_2, T_3, \ldots, T_{k+1}$ so that for all $i = 1, 2, 3, \ldots, k$, T_i beats T_{i+1}.

Consider any round-robin tournament with $k + 1$ teams. Pick one and call it T'. Temporarily remove T' and consider the remaining k teams. Since each of these teams plays each other team exactly once, the games played by these k teams form a round-robin tournament. It follows by inductive hypothesis that these k teams may be labeled $T_1, T_2, T_3, \ldots, T_k$ where T_i beats T_{i+1} for all $i = 1, 2, 3, \ldots, k - 1$.

Case 1 (T' beats T_1): In this case, relabel each T_i to be T_{i+1}, and let $T_1 = T'$. Then T_1 beats the newly labeled T_2 (because T' beats the old T_1), and T_i beats T_{i+1} for all $i = 2, 3, \ldots, k$ (by inductive hypothesis).

Case 2 (T' loses to $T_1, , T_2, T_3, \ldots, T_m$ and beats T_{m+1} where $1 \le m \le k - 1$): In this case, relabel teams $T_{m+1}, T_{m+2}, \ldots, T_k$ to be $T_{m+2}, T_{m+3}, \ldots, T_{k+1}$ and let $T_{m+1} = T'$. Then for each i with $1 \le i \le m - 1$, T_i beats T_{i+1} (by inductive hypothesis), T_m beats T_{m+1} (because T_m beats T'), T_{m+1} beats T_{m+2} (because T' beats the old T_{m+1}), and for each i with $m + 2 \le i \le k$, T_i beats T_{i+1} (by inductive hypothesis).

Case 3 (T' loses to T_i for all $i = 1, 2, \ldots, k$): In this case, let $T_{k+1} = T'$. Then for all $i = 1, 2, \ldots, k - 1$, T_i beats T_{i+1} (by inductive hypothesis) and T_k beats T_{k+1} (because T_k beats T').

Thus in all three cases the teams may be relabeled in the way specified *[as was to be shown]*.

39. <u>Proof (by mathematical induction):</u> Let the property $P(n)$ be the sentence

> The interior angles of any n-sided convex polygon add up to $180(n - 2)$ degrees.

We will prove that $P(n)$ is true for all integers $n \ge 3$.

Show that $P(3)$ is true: $P(3)$ is true because any convex 3-sided polygon is a triangle, the sum of the interior angles of any triangle is 180 degrees, and $180(3 - 2) = 180$. So the angles of any 3-sided convex polygon add up to $180(3 - 2)$ degrees.

Show that for all integers $k \ge 3$, if $P(k)$ is true then $P(k + 1)$ is true: Let k be any integer with $k \ge 3$ and suppose that

> The interior angles of any k-sided convex polygon add up to $180(k - 2)$ degrees. $\leftarrow$ inductive hypothesis

We must show that

> The interior angles of any $(k + 1)$-sided convex polygon add up to $180((k + 1) - 2) = 180(k - 1)$ degrees.

Let p be any $(k + 1)$-sided convex polygon. Label the vertices of p as $v_1, v_2, v_3, \ldots, v_k, v_{k+1}$, and draw a straight line from v_1 to v_3. Because the angles at v_1, v_2, and v_3 are all less than 180 degrees, this line lies entirely inside the polygon. Thus polygon p is split in two pieces: (1) the polygon p' obtained from p by using all of its vertices except v_2, and (2) triangle t with vertices v_1, v_2, v_3. This situation is illustrated in the diagram below.

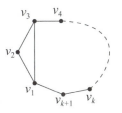

Note that p' has k vertices, and it is convex because the sizes of the angles at v_1 and v_3 in p' are less than their sizes in p. Because p' is a convex polygon with k vertices, by inductive hypothesis, the sum of its interior angles is $180(k-2)$ degrees. Now polygon p is obtained by joining p' and t, and since the sum of the interior angles in t is 180 degrees,

$$
\begin{aligned}
\text{the sum of the interior angles in } p \quad &= \quad \text{the sum of the interior angles in } p' \\
&\qquad + \text{ the sum of the interior angles in } t \\
&= \quad 180(k-2) \text{ degrees} + 180 \text{ degrees} \\
&= \quad 180(k-2+1) \text{ degrees} \\
&= \quad 180(k-1) \text{ degrees},
\end{aligned}
$$

as was to be shown.

Section 5.4

3. <u>Proof (by strong mathematical induction)</u>: Let the property $P(n)$ be the sentence

$$c_n \text{ is even.}$$

We will prove that $P(n)$ is true for all integers $n \geq 0$.

Show that $P(0)$, $P(1)$, and $P(2)$ are true: By definition of $c_0, c_1, c_2, \ldots$, we have that $c_0 = 2$, $c_1 = 2$, and $c_2 = 6$ and 2, 2, and 6 are all even. So $P(0)$, $P(1)$, and $P(2)$ are all true.

Show that if $k \geq 2$ and $P(i)$ is true for all integers i from 0 through k, then $P(k+1)$ is true: Let k be any integer with $k \geq 2$, and suppose

$$c_i \text{ is even for all integers } i \text{ with } 0 \leq i \leq k \qquad \leftarrow \text{inductive hypothesis}$$

We must show that

$$c_{k+1} \text{ is even.}$$

But by definition of $c_0, c_1, c_2, \ldots, c_{k+1} = 3c_{k-2}$. Since $k \geq 2$, we have that $0 \leq k-2 \leq k$, and so, by inductive hypothesis, c_{k-2} is even. But the product of an even integer with any integer is even *[properties 1 and 4 of Example 4.2.3]*, and hence $3c_{k-2}$, which equals c_{k+1}, is also even *[as was to be shown]*.

[Since both the basis and the inductive steps have been proved, we conclude that $P(n)$ is true for all integers $n \geq 0$.]

6. <u>Proof (by strong mathematical induction)</u>: Let the property $P(n)$ be the equation

$$f_n = 3 \cdot 2^n + 2 \cdot 5^n.$$

We will prove that $P(n)$ is true for all integers $n \geq 0$.

Show that $P(0)$ and $P(1)$ are true: By definition of $f_0, f_1, f_2, \ldots$, we have that $f_0 = 5$ and $f_1 = 16$. Since $3 \cdot 2^0 + 2 \cdot 5^0 = 3 + 2 = 5$ and $3 \cdot 2^1 + 2 \cdot 5^1 = 6 + 10 = 16$, $P(0)$ and $P(1)$ are both true.

Show that if $k \geq 1$ and $P(i)$ is true for all integers i from 0 through k, then $P(k+1)$ is true: Let k be any integer with $k \geq 1$, and suppose

$$f_i = 3 \cdot 2^i + 2 \cdot 5^i \text{ for all integers } i \text{ with } 0 \leq i \leq k. \qquad \leftarrow \text{inductive hypothesis}$$

We must show that

$$f_{k+1} = 3 \cdot 2^{k+1} + 2 \cdot 5^{k+1}.$$

But

$$
\begin{aligned}
f_{k+1} &= 7f_k - 10f_{k-1} && \text{by definition of } f_0, f_1, f_2, \ldots \\
&= 7(3 \cdot 2^k + 2 \cdot 5^k) - 10(3 \cdot 2^{k-1} + 2 \cdot 5^{k-1}) && \text{by inductive hypothesis} \\
&= 7(6 \cdot 2^{k-1} + 10 \cdot 5^{k-1}) - 10(3 \cdot 2^{k-1} + 2 \cdot 5^{k-1}) && \text{because } 2^k = 2 \cdot 2^{k-1} \text{ and } 5^k = 5 \cdot 5^{k-1} \\
&= (42 \cdot 2^{k-1} + 70 \cdot 5^{k-1}) - (30 \cdot 2^{k-1} + 20 \cdot 5^{k-1}) \\
&= (42 - 30) \cdot 2^{k-1} + (70 - 20) \cdot 5^{k-1} \\
&= 12 \cdot 2^{k-1} + 50 \cdot 5^{k-1} \\
&= 3 \cdot 2^2 \cdot 2^{k-1} + 2 \cdot 5^2 \cdot 5^{k-1} \\
&= 3 \cdot 2^{k+1} + 2 \cdot 5^{k+1} && \text{by algebra,}
\end{aligned}
$$

[as was to be shown].

[Since both the basis and the inductive steps have been proved, we conclude that $P(n)$ is true for all integers $n \geq 0$.]

9. <u>Proof (by strong mathematical induction)</u>: Let the property $P(n)$ be the inequality

$$
a_n \leq \left(\frac{7}{4}\right)^n .
$$

We will prove that $P(n)$ is true for all integers $n \geq 1$.

Show that $P(1)$ and $P(2)$ are true: By definition of $a_1, a_2, a_3, \ldots$, we have that $a_1 = 1$ and $a_2 = 3$. But

$$
\frac{7}{4} > 1 \quad \text{and} \quad \left(\frac{7}{4}\right)^2 = \frac{49}{16} = 3\frac{1}{16} > 3
$$

So $a_1 \leq \frac{7}{4}$ and $a_2 \leq \left(\frac{7}{4}\right)^2$, and thus $P(1)$ and $P(2)$ are both true.

Show that if $k \geq 2$ and $P(i)$ is true for all integers i from 1 through k, then $P(k+1)$ is true: Let k be any integer with $k \geq 2$, and suppose

$$
a_i \leq \left(\frac{7}{4}\right)^i \text{ for all integers } i \text{ with } 0 \leq i \leq k. \qquad \leftarrow \text{ inductive hypothesis}
$$

We must show that
$$
a_{k+1} \leq \left(\frac{7}{4}\right)^{k+1} .
$$

Since $k \geq 2$,

$$
\begin{aligned}
a_{k+1} &= a_k + a_{k-1} && \text{by definition of } a_1, a_2, a_3, \ldots \\
\Rightarrow \quad a_k + a_{k-1} &\leq \left(\frac{7}{4}\right)^k + \left(\frac{7}{4}\right)^{k-1} && \text{by inductive hypothesis} \\
\Rightarrow \quad a_k + a_{k-1} &\leq \left(\frac{7}{4}\right)^{k-1}\left(\frac{7}{4} + 1\right) && \text{by factoring out } \left(\frac{7}{4}\right)^{k-1} \\
\Rightarrow \quad a_k + a_{k-1} &\leq \left(\frac{7}{4}\right)^{k-1}\left(\frac{11}{4}\right) && \text{by adding } \frac{7}{4} \text{ and } 1 \\
\Rightarrow \quad a_k + a_{k-1} &\leq \left(\frac{7}{4}\right)^{k-1}\left(\frac{44}{16}\right) && \text{by multiplying numerator and denominator of } \frac{11}{4} \text{ by } 4 \\
\Rightarrow \quad a_k + a_{k-1} &\leq \left(\frac{7}{4}\right)^{k-1}\left(\frac{49}{16}\right) && \text{because } \frac{44}{16} < \frac{49}{16} \\
\Rightarrow \quad a_k + a_{k-1} &\leq \left(\frac{7}{4}\right)^{k-1}\left(\frac{7}{4}\right)^2 && \text{because } \left(\frac{49}{16}\right) = \left(\frac{7}{4}\right)^2 \\
\Rightarrow \quad a_k + a_{k-1} &\leq \left(\frac{7}{4}\right)^{k+1} && \text{by a law of exponents.}
\end{aligned}
$$

Thus $a_{k+1} \leq \left(\frac{7}{4}\right)^{k+1}$ *[as was to be shown].*

12. *Note*: This problem can be solved with ordinary mathematical induction.

    Proof (by mathematical induction): Let the property $P(n)$ be the sentence

    > Given any sequence of n cans of gasoline, deposited around a circular track in such a way that the total amount of gasoline is enough for a car to make one complete circuit of the track, it is possible to find an inital location for the car so that it will be able to traverse the entire track by using the various amounts of gasoline in the cans that it encounters along the way.

    We will prove that $P(n)$ is true for all integers $n \geq 1$.

    Show that $P(1)$ is true: When there is just one can, the car should be placed next to it. By hypothesis, the can contains enough gasoline to enable the car to make one complete circuit of the track. Hence $P(1)$ is true.

    Show that for all integers $k \geq 1$, if $P(k)$ is true then $P(k+1)$ is true: Let k be any integer with $k \geq 1$, and suppose that

    > For all integers i with $1 \leq i \leq k$, given any sequence of n cans of gasoline, deposited around a circular track in such a way that the total amount of gasoline is enough for a car to make one complete circuit of the track, it is possible to find an inital location for the car so that it will be able to traverse the entire track by using the various amounts of gasoline in the cans that it encounters along the way. $\leftarrow$ inductive hypothesis

    We must show that

    > Given any sequence of $k+1$ cans of gasoline, deposited around a circular track in such a way that the total amount of gasoline is enough for a car to make one complete circuit of the track, it is possible to find an inital location for the car so that it will be able to traverse the entire track by using the various amounts of gasoline in the cans that it encounters along the way.

    Now, because the total amount of gasoline in all $k+1$ cans is enough for a car to make one complete circuit of the track, there must be at least one can, call it C, that contains enough gasoline to enable the car to reach the next can, say D, in the direction of travel along the track. Imagine pouring all the gasoline from D into C. The result would be k cans deposited around the track in such a way that the total amount of gasoline would be enough for a car to make one complete circuit of the track. By inductive hypothesis, it is possible to find an initial location for the car so that it could traverse the entire track by using the various amounts of gasoline in the cans that it encounters along the way. Use that location as the starting point for the car. When the car reaches can C, the amount of gasoline in C is enough to enable it to reach can D, and once the car reaches D, the additional amount of gasoline in D enables it to complete the circuit. *[This is what was to be shown.]*

15. *Note*: This solution makes free use of the properties from Chapter 4 about sums and differences for two even and odd integers.

    Proof (by strong mathematical induction): Let the property $P(n)$ be the sentence

    > Any sum of n even integers is even.

    We will prove that $P(n)$ is true for all integers $n \geq 2$.

    Show that $P(2)$ is true: $P(2)$ is true because any sum of two even integers is even.

    Show that if $k \geq 2$ and $P(i)$ is true for all integers i from 2 through k, then $P(k+1)$ is true: Let k be any integer with $k \geq 2$, and suppose that

    > For all integers i from 2 through k, any sum of i even integers is even. $\leftarrow$ inductive hypothesis

We must show that

$$\text{any sum of } k+1 \text{ even integers is even.}$$

Consider any sum S of $k+1$ even integers. Some addition is the final one that is used to obtain S. Thus there are integers A and B such that $S = A + B$, A is a sum of r even integers, and B is a sum of $(k+1) - r$ even integers, where $1 \leq r \leq k$ and $1 \leq (k+1) - r \leq k$.

Case 1 (both $2 \leq r \leq k$ and $2 \leq k+1) - r \leq k$): In this case, by inductive hypothesis, both A and B are even, and hence $S = A + B$ is even.

Case 2 ($r = 1$ or $(k+1) - r = 1$): In this case, since $A + B = B + A$, we may assume without loss of generality that $r = 1$. Then A is a single even integer, and, since $2 \leq (k+1) - r \leq k$, B is even by inductive hypothesis. Hence $S = A + B$ is even.

Conclusion: It follows from cases 1 and 2 that any sum of $k+1$ even integers is even *[as was to be shown]*.

18. <u>Conjecture</u>: For all integers $n \geq 0$, the units digit of 9^n is 1 if n is even and is 9 if n is odd.

 <u>Proof (by strong mathematical induction)</u>: Let the property $P(n)$ be the sentence

$$\text{The units digit of } 9^n \text{ is 1 if } n \text{ is even and is 9 if } n \text{ is odd.}$$

 We will prove that $P(n)$ is true for all integers $n \geq 0$.

 Show that $P(0)$ and $P(1)$ are true: $P(0)$ is true because 0 is even and the units digit of $9^0 = 1$. $P(1)$ is true because 1 is odd and the units digit of $9^1 = 9$.

 Show that if $k \geq 1$ and $P(i)$ is true for all integers i from 0 through k, then $P(k+1)$ is true: Let k be any integer with $k \geq 1$, and suppose that

$$\text{The units digit of } 9^n \text{ is 1 if } n \text{ is even and is 9 if } n \text{ is odd.} \qquad \leftarrow \text{inductive hypothesis}$$

 We must show that

$$\text{The units digit of } 9^{k+1} \text{ is 1 if } k+1 \text{ is even and is 9 if } k+1 \text{ is odd.}$$

 Case 1 ($k+1$ is even): In this case k is odd, and so, by inductive hypothesis, the units digit of 9^k is 9. This implies that there is an integer a so that $9^k = 10a + 9$, and hence

$$
\begin{aligned}
9^{k+1} &= 9^1 \cdot 9^k && \text{by algebra (a law of exponents)} \\
&= 9(10a + 9) && \text{by substitution} \\
&= 90a + 81 \\
&= 90a + 80 + 1 \\
&= 10(9a + 8) + 1 && \text{by algebra.}
\end{aligned}
$$

 Because $9a + 8$ is an integer, it follows that the units digit of 9^{k+1} is 1.

 Case 2 ($k+1$ is odd): In this case k is even, and so, by inductive hypothesis, the units digit of 9^{k-1} is 1. This implies that there is an integer a so that $9^k = 10a + 1$, and hence

$$
\begin{aligned}
9^{k+1} &= 9^1 \cdot 9^k && \text{by algebra (a law of exponents)} \\
&= 9(10a + 1) && \text{by substitution} \\
&= 90a + 9 \\
&= 10(9a) + 9 && \text{by algebra.}
\end{aligned}
$$

 Because $9a$ is an integer, it follows that the units digit of 9^{k+1} is 9.

 Hence in both cases the units digit of 9^{k+1} is as specified *[as was to be shown]*.

21. <u>Proof by contradiction</u>: Suppose not. That is, suppose that there exists an integer that is greater than 1, that is not prime, and that is not a product of primes. *[We will show that this supposition leads to a contradiction.]*

    Let S be the set of all integers that are greater than 1, are not prime, and are not a product of primes. That is,

    $$S = \{n \in \mathbf{Z} \mid n > 1, n \text{ is not prime, and } n \text{ is not a product of primes } \}.$$

    Then, by supposition, S has one or more elements. By the well-ordering principle for the integers, S has a least element; call it m. Then m is greater than 1, is not prime, and is not a product of primes.

    Now because m is greater than 1 and is not prime, $m = rs$ for some integers r and s with $1 < r < m$ and $1 < s < m$. Also, because both r and s are less than m, which is the least element of S, neither r nor s is in S. Thus both r and s are either prime or products of primes.

    But this implies that m is a product of primes because m is a product of r and s. Thus m is not in S. So m is in S and m is not in S, which is a contradiction. *[Hence the supposition is false, and so every integer greater than 1 is either prime or a product of primes.]*

27. Suppose $P(n)$ is a property that is defined for integers n and suppose the following statement can be proved using strong mathematical induction:

    $$P(n) \text{ is true for all integers } n \geq a.$$

    Then for some integer $b \geq a$ the following two statements are true:

    1. $P(a), P(a + 1), P(a + 2), \ldots, P(b)$ are all true.

    2. For any integer $k \geq b$, if $P(i)$ is true for all integers i from a through k, then $P(k + 1)$ is true.

    We will show that we can reach the conclusion that $P(n)$ is true for all integers $n \geq a$ using ordinary mathematical induction.

    <u>Proof by mathematical induction</u>: Let $Q(n)$ be the property

    $$P(j) \text{ is true for all integers } j \text{ from } a \text{ through } n.$$

    Show that $Q(b)$ is true: For $n = b$, the property is "$P(j)$ is true for all integers j with $a \leq j \leq b$." But this is true by (1) above.

    Show that for all integers $k \geq b$, if $Q(k)$ is true then $Q(k + 1)$ is true: Let k be any integer with $k \geq b$, and suppose that $Q(k)$ is true. In other words, suppose that

    $$P(j) \text{ is true for all integers } j \text{ from } a \text{ through } k. \quad \leftarrow \text{ inductive hypothesis}$$

    We must show that $Q(k + 1)$ is true. In other words, we must show that

    $$P(j) \text{ is true for all integers } j \text{ from } a \text{ through } k + 1.$$

    Since, by inductive hypothesis, $P(j)$ is true for all integers j from a through k, it follows from (2) above that $P(k+1)$ is also true. Hence $P(j)$ is true for all integers j from a through $k+1$, *[as was to be shown]*.

    It follows by the principle of ordinary mathematical induction that $P(j)$ is true for all integers j from a through n for all integers $n \geq b$. From this and from (1) above, we conclude that $P(n)$ is true for all integers $n \geq a$.

30. <u>Theorem</u>: Given any nonnegative integer n and any positive integer d, there exist integers q and r such that $n = dq + r$ and $0 \le r < d$.

<u>Proof by (ordinary) mathematical induction</u>: Let a nonnegative integer d be given, and let the property $P(n)$ be the sentence

$$\text{There exist integers } q \text{ and } r \text{ such that} \qquad \leftarrow P(n)$$
$$n = dq + r \quad \text{and} \quad 0 \le r < d.$$

We will prove that $P(n)$ is true for all integers $n \ge 0$.

Show that $P(0)$ is true: We must show that there exist integers q and r such that

$$0 = dq + r \quad \text{and} \quad 0 \le r < d.$$

Let $q = r = 0$. Then

$$0 = d \cdot 0 + 0 \quad \text{and} \quad 0 \le 0 < d.$$

Hence $P(0)$ is true.

Show that for all integers $k \ge 0$, if $P(k)$ is true, then $P(k + 1)$ is true: Let k be any integer with $k \ge 0$, and suppose that

$$\text{There exist integers } q' \text{ and } r' \text{ such that} \qquad \leftarrow \quad P(k)$$
$$k = dq' + r' \quad \text{and} \quad 0 \le r' < d. \qquad\qquad\qquad \text{inductive hypothesis}$$

We must show that

$$\text{There exist integers } q \text{ and } r \text{ such that} \qquad \leftarrow P(k+1)$$
$$k + 1 = dq + r \quad \text{and} \quad 0 \le r < d.$$

Adding 1 to both sides of the equation in the inductive hypothesis gives

$$k + 1 = (dq' + r') + 1.$$

Note that since r' is an integer and $0 \le r' < d$, then either $r' < d - 1$ or $r' = d - 1$.

Case 1 ($r' < d - 1$): In this case

$$k + 1 = (dq' + r') + 1 = dq' + (r' + 1).$$

Let $q = q'$ and $r = r' + 1$. Then, by substitution,

$$k + 1 = dq + r.$$

Since

$$r' < d - 1 \quad \text{then} \quad r = r' + 1 < d,$$

and since

$$r = r' + 1 \quad \text{and} \quad r' \ge 0 \quad \text{then} \quad r \ge 0.$$

Hence

$$0 \le r < d.$$

Case 2 ($r' = d - 1$): In this case

$$\begin{aligned} k + 1 &= (dq' + r') + 1 \\ &= dq' + (r' + 1) \\ &= dq' + ((d - 1) + 1) \\ &= dq' + d = d(q' + 1). \end{aligned}$$

Let $q = q' + 1$ and $r = 0$. Then by substitution

$$k + 1 = dq + r,$$

and since

$$r = 0 \quad \text{and} \quad d > 0,$$

then

$$0 \leq r < d.$$

Thus in either case there exist integers q and r such that $k + 1 = dq + r$ and $0 \leq r < d$ [as was to be shown].

Section 5.5

9. Proof

   I. Basis Property: $I(0)$ is the statement

   both a and A are even integers or both are odd integers and, in either case, $a \geq -1$.

   According to the pre-condition this statement is true.

   II. Inductive Property: Suppose k is any nonnegative integer such that $G \wedge I(k)$ is true before an iteration of the loop. Then when execution comes to the top of the loop, $a_{\text{old}} > 0$ and

   both a_{old} and A are even integers or both are odd integers and, in either case, $a_{\text{old}} \geq -1$.

   Execution of statement 1 sets a_{new} equal to $a_{\text{old}} - 2$. Hence a_{new} has the same parity as a_{old} which is the same as A. Also since $a_{\text{old}} > 0$, then

   $$a_{\text{new}} = a_{\text{old}} - 2 > 0 - 2 = -2.$$

   But since $a_{\text{new}} > -2$ and since a_{new} is an integer, $a_{\text{new}} \geq -1$. Hence after the loop iteration, $I(k + 1)$ is true.

   III. Eventual Falsity of Guard: The guard G is the condition $a > 0$. After each iteration of the loop,
   $$a_{\text{new}} = a_{\text{old}} - 2 < a_{\text{old}},$$

   and so successive iterations of the loop give a strictly decreasing sequence of integer values of a which eventually becomes less than or equal to zero, at which point G becomes false.

   IV. Correctness of the Post-Condition: Suppose that N is the least number of iterations after which G is false and $I(N)$ is true. Then (since G is false) $a \leq 0$ and (since $I(N)$ is true) both a and A are even integers or both are odd integers, and $a \geq -1$. Putting the inequalities together gives
   $$-1 \leq a \leq 0,$$

   and so since a is an integer, $a = -1$ or $a = 0$. Since a and A have the same parity, then, $a = 0$ if A is even and $a = -1$ if A is odd. This is the post-condition.

12. *a.* Suppose the following condition is satisfied before entry to the loop: "there exist integers u, v, s, and t such that $a = uA + vB$ and $b = sA + tB$." Then

   $$a_{\text{old}} = u_{\text{old}}A + v_{\text{old}}B \quad \text{and} \quad b_{\text{old}} = s_{\text{old}}A + t_{\text{old}}B,$$

for some integers u_{old}, v_{old}, s_{old}, and t_{old}. Observe that $b_{\text{new}} = r_{\text{new}} = a_{\text{old}} \bmod b_{\text{old}}$. So by the quotient-remainder theorem, there exists a unique integer q_{new} with $a_{\text{old}} = b_{\text{old}} \cdot q_{\text{new}} + r_{\text{new}} = b_{\text{old}} \cdot q_{\text{new}} + b_{\text{new}}$. Solving for b_{new} gives

$$\begin{aligned}
b_{\text{new}} &= a_{\text{old}} - b_{\text{old}} \cdot q_{\text{new}} \\
&= (u_{\text{old}} A + v_{\text{old}} B) - (s_{\text{old}} A + t_{\text{old}} B) q_{\text{new}} \\
&= (u_{\text{old}} - s_{\text{old}} q_{\text{new}}) A + (v_{\text{old}} - t_{\text{old}} q_{\text{new}}) B.
\end{aligned}$$

Therefore, let

$$s_{\text{new}} = u_{\text{old}} - s_{\text{old}} q_{\text{new}} \quad \text{and} \quad t_{\text{new}} = v_{\text{old}} - t_{\text{old}} q_{\text{new}}.$$

Also since

$$a_{\text{new}} = b_{\text{old}} = s_{\text{old}} A + t_{\text{old}} B,$$

let

$$u_{\text{new}} = s_{\text{old}} \quad \text{and} \quad v_{\text{new}} = t_{\text{old}}.$$

Hence

$$a_{\text{new}} = u_{\text{new}} \cdot A + v_{\text{new}} \cdot B \quad \text{and} \quad b_{\text{new}} = s_{\text{new}} \cdot A + t_{\text{new}} \cdot B,$$

and so the condition is true after each iteration of the loop and hence after exit from the loop.

b. Initially $a = A$ and $b = B$. Let $u = 1$, $v = 0$, $s = 0$, and $t = 1$. Then before the first iteration of the loop,

$$a = uA + vB \quad \text{and} \quad b = sA + tB,$$

as was to be shown.

c. By part (b) there exist integers u, v, s, and t such that before the first iteration of the loop,

$$a = uA + vB \quad \text{and} \quad b = sA + tB.$$

So by part (a), after each subsequent iteration of the loop, there exist integers u, v, s, and t such that

$$a = uA + vB \quad \text{and} \quad b = sA + tB.$$

Now after the final iteration of the **while** loop in the Euclidean algorithm, the variable *gcd* is given the current value of a. (See page 224.) But by the correctness proof for the Euclidean algorithm, $gcd = \gcd(A, B)$. Hence there exist integers u and v such that

$$\gcd(A, B) = uA + vB.$$

d. The method discussed in part (a) gives the following formulas for u, v, s, and t:

$$u_{\text{new}} = s_{\text{old}}, \qquad v_{\text{new}} = t_{\text{old}}, \qquad s_{\text{new}} = u_{\text{old}} - s_{\text{old}} q_{\text{new}}, \qquad \text{and} \qquad t_{\text{new}} = v_{\text{old}} - t_{\text{old}} q_{\text{new}},$$

where in each iteration q_{new} is the quotient obtained by dividing a_{old} by b_{old}. The trace table below shows the values of a, b, r, q, gcd, and u, v, s, and t for the iterations of the **while** loop from the Euclidean algorithm. By part (b) the initial values of u, v, s, and t are $u = 1$, $v = 0$, $s = 0$, and $t = 1$.

r		18	12	6	0
q		2	8	1	2
a	330	156	18	12	6
b	156	18	12	6	0
gcd					6
u	1	0	1	-8	9
v	0	1	-2	17	-19
s	0	1	-8	9	-26
t	1	-2	17	-19	55

Since the final values of gcd, u, and v are 6, 9 and -19 and since $A = 330$ and $B = 156$, we have $\gcd(330, 156) = 6 = 330u + 156v = 330 \cdot 9 + 156 \cdot (-19)$, which is true.

Section 5.6

6. $t_0 = -1$, $t_1 = 2$, $t_2 = t_1 + 2 \cdot t_0 = 2 + 2 \cdot (-1) = 0$, $t_3 = t_2 + 2 \cdot t_1 = 0 + 2 \cdot 2 = 4$

12. For all integers $n \geq 0$, $s_n = \frac{(-1)^n}{n!}$. Thus for any integer k with $k \geq 1$,

$$s_k = \frac{(-1)^k}{k!} \quad \text{and} \quad s_{k-1} = \frac{(-1)^{k-1}}{(k-1)!}.$$

It follows that for any integer k with $k \geq 1$,

$$
\begin{aligned}
\frac{-s_{k-1}}{k} &= \frac{-\frac{(-1)^{k-1}}{(k-1)!}}{k} && \text{by substitution} \\
&= \frac{-(-1)^{k-1}}{k(k-1)!} \\
&= \frac{(-1)^k}{k!} \\
&= s_k && \text{by algebra.}
\end{aligned}
$$

Thus $s_k = \dfrac{-s_{k-1}}{k}$.

15. **Proof** : Let n be an integer with $n \geq 1$. Then

$$
\begin{aligned}
\frac{1}{4n+2}\binom{2n+2}{n+1} &= \left(\frac{1}{2(2n+1)}\right)\left(\frac{(2n+2)!}{(n+1)!((2n+2)-(n+1))!}\right) \\
&= \left(\frac{1}{2(2n+1)}\right)\left(\frac{(2n+2)!}{(n+1)!(n+1)!}\right) \\
&= \left(\frac{1}{2(2n+1)}\right)\left(\frac{(2n+2)(2n+1)(2n)!}{(n+1)\cdot n! \cdot (n+1)\cdot n!}\right) \\
&= \frac{1}{2}\left(\frac{2(n+1)}{(n+1)\cdot(n+1)}\right)\left(\frac{(2n)!}{n!\cdot n!}\right) \\
&= \frac{1}{n+1}\binom{2n}{n} \\
&= C_n.
\end{aligned}
$$

Thus $C_n = \dfrac{1}{4n+2}\dbinom{2n+2}{n+1}$.

18. *a.* $b_1 = 1$, $b_2 = 1 + 1 + 1 + 1 = 4$, $b_3 = 4 + 4 + 1 + 4 = 13$

c. Note that it takes just as many moves to move a stack of disks from the middle pole to an outer pole as from an outer pole to the middle pole: the moves are the same except that their order and direction are reversed. For all integers $k \geq 2$,

$$
\begin{aligned}
b_k &= a_{k-1} && \text{(moves to transfer the top } k-1 \text{ disks from pole } A \text{ to pole } C\text{)} \\
&\quad +1 && \text{(move to transfer the bottom disk from pole } A \text{ to pole } B\text{)} \\
&\quad +b_{k-1} && \text{(moves to transfer the top } k-1 \text{ disks from pole } C \text{ to pole } B\text{).} \\
&= a_{k-1} + 1 + b_{k-1}.
\end{aligned}
$$

d. One way to transfer a tower of k disks from pole A to pole B is first to transfer the top $k-1$ disks from pole A to pole B *[this requires b_{k-1} moves]*, then transfer the top $k-1$ disks from pole B to pole C *[this also requires b_{k-1} moves]*, then transfer the bottom disk from pole A to pole B *[this requires one move]*, and finally transfer the top $k-1$ disks from pole C to pole B *[this again requires b_{k-1} moves]*. This sequence of steps need not necessarily, however, result in a minimum number of moves. Therefore, at this point, all we can say for sure is that for all integers $k \geq 2$,

$$b_k \leq b_{k-1} + b_{k-1} + 1 + b_{k-1} = 3b_{k-1} + 1.$$

e. <u>Proof (by mathematical induction)</u>: Let the property $P(k)$ be the equation

$$b_k = 3b_{k-1} + 1.$$

Show that $P(2)$ is true: The property is true for $k = 2$ because for $k = 2$ the left-hand side is 4 (by part (a)) and the right-hand side is $3 \cdot 1 + 1 = 4$ also.

Show that for all integers $i \geq 2$, if $P(i)$ is true then $P(i+1)$ is true: Let i be any integer with $i \geq 2$, and suppose that

$$b_i = 3b_{i-1} + 1. \qquad \leftarrow \quad \begin{array}{c} P(i) \\ \text{inductive hypothesis} \end{array}$$

We must show that

$$b_{i+1} = 3b_i + 1. \qquad \leftarrow \quad P(i+1)$$

But the left-hand side of $P(i+1)$ is

$$
\begin{aligned}
b_{i+1} &= a_i + 1 + b_i & &\text{by part (c)} \\
&= a_i + 1 + 3b_{i-1} + 1 & &\text{by inductive hypothesis} \\
&= (3a_{i-1} + 2) + 1 + 3b_{i-1} + 1 & &\text{by exercise 17 (c)} \\
&= 3a_{i-1} + 3 + 3b_{i-1} + 1 & & \\
&= 3(a_{i-1} + 1 + b_{i-1}) + 1 & &\text{by algebra} \\
&= 3b_i + 1 & &\text{by part (c) of this exercise,}
\end{aligned}
$$

which is the right-hand side of $P(i+1)$ *[as was to be shown]*.

21. *a.* $t_1 = 2, \quad t_2 = 2 + 2 + 2 = 6$

   c. For all integers $k \geq 2$,

$$
\begin{aligned}
t_k &= t_{k-1} & &\text{(moves to transfer the top } 2k-2 \text{ disks from pole } A \text{ to pole } B\text{)} \\
&\quad +2 & &\text{(moves to transfer the bottom two disks from pole } A \text{ to pole } C\text{)} \\
&\quad +t_{k-1} & &\text{(moves to transfer the top } 2k-2 \text{ disks from pole } B \text{ to pole } C\text{)} \\
&= 2t_{k-1} + 2.
\end{aligned}
$$

Note that transferring the stack of $2k$ disks from pole A to pole C requires at least two transfers of the top $2(k-1)$ disks: one to transfer them off the bottom two disks to free the bottom disks so that they can be moved to pole C and another to transfer the top $2(k-1)$ disks back on top of the bottom two disks. Thus at least $2t_{k-1}$ moves are needed to effect these two transfers. Two more moves are needed to transfer the bottom two disks from pole A to pole C, and this transfer cannot be effected in fewer than two moves. It follows that the sequence of moves indicated in the description of the equation above is, in fact, minimal.

24. $F_{13} = F_{12} + F_{11} = 233 + 144 = 377, \quad F_{14} = F_{13} + F_{12} = 377 + 233 = 610$

30. **Proof (by mathematical induction):** Let the property $P(n)$ be the equation

$$F_{n+2}F_n - F_{n+1}^2 = (-1)^n. \qquad \leftarrow P(n)$$

Show that $P(0)$ is true: The left-hand side of $P(0)$ is $F_{0+2}F_0 - F_1^2 = 2 \cdot 1 - 1^2 = 1$, and the right-hand side is $(-1)^0 = 1$ also. So $P(0)$ is true.

Show that for all integers $k \geq 0$, if $P(k)$ is true then $P(k+1)$ is true: Let k be any integer with $k \geq 0$, and suppose that

$$F_{k+2}F_k - F_{k+1}^2 = (-1)^k. \qquad \leftarrow \begin{array}{l} P(k) \\ \text{inductive hypothesis} \end{array}$$

We must show that

$$F_{k+3}F_{k+1} - F_{k+2}^2 = (-1)^{k+1}. \qquad \leftarrow P(k+1)$$

But by inductive hypothesis,

$$F_{k+1}^2 = F_{k+2}F_k - (-1)^k = F_{k+2}F_k + (-1)^{k+1}. \quad (*)$$

Hence,

$$
\begin{aligned}
F_{k+3}&F_{k+1} - F_{k+2}^2 \\
&= (F_{k+1} + F_{k+2})F_{k+1} - F_{k+2}^2 && \text{by definition of the Fibonacci sequence} \\
&= F_{k+1}^2 + F_{k+2}F_{k+1} - F_{k+2}^2 \\
&= F_{k+2}F_k + (-1)^{k+1} + F_{k+2}F_{k+1} - F_{k+2}^2 && \text{by substitution from equation } (*) \\
&= F_{k+2}(F_k + F_{k+1} - F_{k+2}) + (-1)^{k+1} && \text{by factoring out } F_{k+2} \\
&= F_{k+2}(F_{k+2} - F_{k+2}) + (-1)^{k+1} && \text{by definition of the Fibonacci sequence} \\
&= F_{k+2} \cdot 0 + (-1)^{k+1} \\
&= (-1)^{k+1}.
\end{aligned}
$$

36. Let $L = \lim_{n \to \infty} x_n$. By definition of $x_0, x_1, x_2, \ldots$ and by the continuity of the square root function,

$$L = \lim_{n \to \infty} x_n = \lim_{n \to \infty} \sqrt{2 + x_{n-1}} = \sqrt{2 + \lim_{n \to \infty} x_{n-1}} = \sqrt{2 + L}.$$

Hence $L^2 = 2 + L$, and so $L^2 - L - 2 = 0$. Factoring gives $(L - 2)(L + 1) = 0$, and so $L = 2$ or $L = -1$. But $L \geq 0$ because each $x_i \geq 0$. Thus $L = 2$.

42. **Proof (by mathematical induction):** Let the property $P(n)$ be the sentence

$$\text{If } a_1, a_2, \ldots, a_n \text{ and } b_1, b_2, \ldots, b_n \text{ are any real} \atop \text{numbers, then } \prod_{i=1}^{n}(a_i b_i) = \left(\prod_{i=1}^{n} a_i\right)\left(\prod_{i=1}^{n} b_i\right). \qquad \leftarrow P(n)$$

Show that $P(1)$ is true: Let a_1 and b_1 be any real numbers. By the recursive definition of product,

$$\prod_{i=1}^{1}(a_i b_i) = a_1 b_1, \prod_{i=1}^{1} a_i = a_1, and \prod_{i=1}^{1} b_i = b_1.$$

Show that for all integers $k \geq 1$, if $P(k)$ is true then $P(k+1)$ is true: Let k be any integer with $k \geq 1$, and suppose that

$$\text{If } a_1, a_2, \ldots, a_k \text{ and } b_1, b_2, \ldots, b_k \text{ are any real} \atop \text{numbers, then } \prod_{i=1}^{k}(a_i b_i) = \left(\prod_{i=1}^{k} a_i\right)\left(\prod_{i=1}^{k} b_i\right). \qquad \leftarrow \begin{array}{l} P(k) \\ \text{inductive hypothesis} \end{array}$$

We must show that

If $a_1, a_2, \ldots, a_{k+1}$ and $b_1, b_2, \ldots, b_{k+1}$ are any real
numbers, then $\prod_{i=1}^{k+1}(a_ib_i) = \left(\prod_{i=1}^{k+1} a_i\right)\left(\prod_{i=1}^{k+1} b_i\right).$ $\leftarrow P(k+1)$

So suppose $a_1, a_2, \ldots, a_{k+1}$ and $b_1, b_2, \ldots, b_{k+1}$ are any real numbers. Then

$\prod_{i=1}^{k+1}(a_ib_i)$

$$= \left(\prod_{i=1}^{k}(a_ib_i)\right)(a_{k+1}b_{k+1}) \qquad \text{by the recursive definition of product}$$

$$= \left(\left(\prod_{i=1}^{k} a_i\right)\left(\prod_{i=1}^{k} b_i\right)\right)(a_{k+1}b_{k+1}) \qquad \text{by substitution from the inductive hypothesis}$$

$$= \left(\left(\prod_{i=1}^{k} a_i\right)a_{k+1}\right)\left(\left(\prod_{i=1}^{k} b_i\right)b_{k+1}\right) \qquad \text{by the associative and commutative laws of algebra}$$

$$= \left(\prod_{i=1}^{k+1} a_i\right)\left(\prod_{i=1}^{k+1} b_i\right) \qquad \text{by the recursive definition of product.}$$

[This is what was to be shown.]

Section 5.7

6. $$d_1 = 2$$
$$d_2 = 2d_1 + 3 = 2 \cdot 2 + 3 = 2^2 + 3$$
$$d_3 = 2d_2 + 3 = 2(2^2 + 3) + 3 = 2^3 + 2 \cdot 3 + 3$$
$$d_4 = 2d_3 + 3 = 2(2^3 + 2 \cdot 3 + 3) + 3 = 2^4 + 2^2 \cdot 3 + 2 \cdot 3 + 3$$
$$d_5 = 2d_4 + 3 = 2(2^4 + 2^2 \cdot 3 + 2 \cdot 3 + 3) + 3 = 2^5 + 2^3 \cdot 3 + 2^2 \cdot 3 + 2 \cdot 3 + 3$$

.
.
.

Guess:
$$\begin{aligned}
d_n &= 2^n + 2^{n-2} \cdot 3 + 2^{n-3} \cdot 3 + \cdots + 2^2 \cdot 3 + 2 \cdot 3 + 3 \\
&= 2^n + 3(2^{n-2} + 2^{n-3} + \cdots + 2^2 + 2 + 1) \\
&= 2^n + 3\left(\frac{2^{(n-2)+1} - 1}{2 - 1}\right) \quad \textit{[by Theorem 5.2.3]} \\
&= 2^n + 3(2^{n-1} - 1) \\
&= 2^{n-1}(2 + 3) - 3 = 5 \cdot 2^{n-1} - 3 \quad \text{for all integers } n \geq 1
\end{aligned}$$

9. $$g_1 = 1$$
$$g_2 = \frac{g_1}{g_1 + 2} = \frac{1}{1 + 2}$$

$$g_3 = \frac{g_2}{g_2 + 2} = \frac{\frac{1}{1+2}}{\frac{1}{1+2} + 2} = \frac{1}{1 + 2(1 + 2)} = \frac{1}{1 + 2 + 2^2}$$

$$g_4 = \frac{g_3}{g_3 + 2} = \frac{\frac{1}{1+2+2^2}}{\frac{1}{1+2+2^2} + 2} = \frac{1}{1 + 2(1 + 2 + 2^2)} = \frac{1}{1 + 2 + 2^2 + 2^3}$$

$$g_5 = \frac{g_4}{g_4 + 2} = \frac{\frac{1}{1+2+2^2+2^3}}{\frac{1}{1+2+2^2+2^3} + 2} = \frac{1}{1 + 2(1 + 2 + 2^2 + 2^3)} = \frac{1}{1 + 2 + 2^2 + 2^3 + 2^4}$$

.
.
.

Guess:
$$\begin{aligned}
g_n &= \frac{1}{1 + 2 + 2^2 + 2^3 + \cdots + 2^{n-1}} \\
&= \frac{1}{2^n - 1} \quad \textit{[by Theorem 5.2.3)]} \quad \text{for all integers } n \geq 1
\end{aligned}$$

15.
$$y_1 = 1$$
$$y_2 = y_1 + 2^2 = 1 + 2^2$$
$$y_3 = y_2 + 3^2 = (1 + 2^2) + 3^2 = 1 + 2^2 + 3^2$$
$$y_4 = y_3 + 4^2 = (1 + 2^2 + 3^2) + 4^2 = 1^2 + 2^2 + 3^2 + 4^2$$

.

.

.

Guess:

$$y_n = 1^2 + 2^2 + 3^2 + \cdots + n^2 = \frac{n(n+1)(2n+1)}{6} \text{ by exercise 10 of Section 5.2}$$

21. **Proof (by mathematical induction)**: Let r be a fixed constant and $a_0, a_1, a_2, \ldots$ a sequence that satisfies the recurrence relation $a_k = ra_{k-1}$ for all integers $k \geq 1$ and the initial condition $a_0 = a$. Let the property $P(n)$ be the equation $a_n = ar^n$.

$$a_n = ar^n.$$

Show that $P(0)$ is true: The right-hand side of $P(0)$ is $ar^0 = a \cdot 1 = a$, which is the left-hand side of $P(0)$. So $P(0)$ is true.

Show that for all integers $k \geq 0$, if $P(k)$ is true then $P(k+1)$ is true: Let k be any integer with $k \geq 0$, and suppose that

$$a_k = ar^k. \quad \leftarrow \quad \begin{array}{c} P(k) \\ \text{inductive hypothesis} \end{array}$$

We must show that

$$a_{k+1} = ar^{k+1}. \quad \leftarrow \quad P(k+1)$$

The left-hand side of $P(k+1)$ is

$$\begin{aligned} a_{k+1} &= ra_k & \text{by definition of } a_0, a_1, a_2, \ldots \\ &= r(ar^k) & \text{by inductive hypothesis} \\ &= ar^{k+1} & \text{by the laws of exponents,} \end{aligned}$$

which is the right-hand side of $P(k+1)$, *[as to be shown]*.

27. *a.* Let the original balance in the account be A dollars, and let A_n be the amount owed in month n assuming the balance is not reduced by making payments during the year. The annual interest rate is 18%, and so the monthly interest rate is $(18/12)\% = 1.5\% = 0.015$. The sequence $A_0, A_1, A_2, \ldots$ satisfies the recurrence relation

$$A_k = A_{k-1} + 0.015A_{k-1} = 1.015A_{k-1}.$$

Thus
$$A_1 = 1.015A_0 = 1.015A$$
$$A_2 = 1.015A_1 = 1.015(1.015A) = (1.015)^2 A$$

$$\vdots$$

$$A_{12} = 1.015A_{11} = 1.015(1.015)^{11}A = (1.015)^{12}A.$$

So the amount owed at the end of the year is $(1.015)^{12}A$. It follows that the APR is

$$\frac{(1.015)^{12}A - A}{A} = \frac{A((1.015)^{12} - 1)}{A} = (1.015)^{12} - 1 \cong 19.6\%.$$

Note: Because $A_k = 1.015A_{k-1}$ for each integer $k \geq 1$, we could have immediately concluded that the sequence is geometric and, therefore, satisfies the equation $A_n = A_0(1.015)^n = A(1.015)^n$.

b. Because the person pays \$150 per month to pay off the loan, the balance at the end of month k is $B_k = 1.015B_{k-1} - 150$. We use iteration to find an explicit formula for $B_0, B_1, B_2, \ldots$.

$$B_0 = 3000$$

$$B_1 = (1.015)B_0 - 150 = 1.015(3000) - 150$$

$$B_2 = (1.015)B_1 - 150 = (1.015)[1.015(3000) - 150] - 150$$

$$= 3000(1.015)^2 - 150(1.015) - 150$$

$$B_3 = (1.015)B_2 - 150 = (1.015)[3000(1.015)^2 - 150(1.015) - 150] - 150$$

$$= 3000(1.015)^3 - 150(1.015)^2 - 150(1.015) - 150$$

$$B_4 = (1.015)B_3 - 150$$

$$= (1.015)[3000(1.015)^3 - 150(1.015)^2 - 150(1.015) - 150] - 150$$

$$= 3000(1.015)^4 - 150(1.015)^3 - 150(1.015)^2 - 150(1.015) - 150$$

$$\vdots$$

Guess: $B_n = 3000(1.015)^n + [150(1.015)^{n-1} - 150(1.015)^{n-2} + \ldots$
$$-150(1.015)^2 - 150(1.015) - 150]$$

$$= 3000(1.015)^n - 150[(1.015)^{n-1} + (1.015)^{n-2} + \cdots + (1.015)^2 + 1.015 + 1]$$

$$= 3000(1.015)^n - 150 \left(\frac{(1.015)^n - 1}{1.015 - 1} \right)$$

$$= (1.015)^n (3000) - \frac{150}{0.015}((1.015)^n - 1)$$

$$= (1.015)^n (3000) - 10000((1.015)^n - 1)$$

$$= (1.015)^n (3000 - 10000) + 10000$$

$$= (-7000)(1.015)^n + 10000$$

So it appears that $B_n = (-7000)(1.015)^n + 10000$. We use mathematical induction to confirm this guess.

Proof (by mathematical induction): : Let $B_0, B_1, B_2, \ldots$. be a sequence that satisfies the recurrence relation $B_k = (1.015)B_{k-1} - 150$ for all integers $k \geq 1$, with initial condition $B_0 = 3000$, and let the property $P(n)$ be the equation

$$B_n = (-7000)(1.015)^n + 10000. \qquad \leftarrow P(n)$$

Show that $P(0)$ is true: The right-hand side of $P(0)$ is $(-7000)(1.015)^0 + 10000 = 3000$, which equals B_0, the left-hand side of $P(0)$. So $P(0)$ is true.

Show that for all integers $k \geq 0$, if $P(k)$ is true then $P(k+1)$ is true: Let k be any integer with $k \geq 0$, and suppose that

$$B_k = (-7000)(1.015)^k + 10000. \qquad \leftarrow \begin{array}{l} P(k) \\ \text{inductive hypothesis} \end{array}$$

We must show that

$$B_{k+1} = (-7000)(1.015)^{k+1} + 10000. \qquad \leftarrow P(k+1)$$

The left-hand side of $P(k+1)$ is

$$
\begin{aligned}
B_{k+1} &= (1.015)B_k - 150 && \text{by definition of } B_0, B_1, B_2, \ldots \\
&= (1.015)[(-7000)(1.015)^k + 10000] - 150 && \text{by substitution from} \\
& && \text{the inductive hypothesis} \\
&= (-7000)(1.015)^{k+1} + 10150 - 150 \\
&= (-7000)(1.015)^{k+1} + 10000 && \text{by the laws of algebra,}
\end{aligned}
$$

which is the right-hand side of $P(k+1)$ [as to be shown].

c. By part (b), $B_n = (-7000)(1.015)^n + 10000$, and so we need to find the value of n for which

$$(-7000)(1.015)^n + 10000 = 0.$$

But this equation holds

$$\Leftrightarrow \qquad 7000(1.015)^n = 10000$$

$$\Leftrightarrow \qquad (1.015)^n = \frac{10000}{7000} = \frac{10}{7}$$

$$\Leftrightarrow \quad \log_{10}(1.015)^n = \log_{10}\left(\frac{10}{7}\right) \qquad \text{by a property of logarithms}$$

$$\Leftrightarrow \quad n\log_{10}(1.015) = \log_{10}\left(\frac{10}{7}\right) \qquad \text{by a property of logarithms}$$

$$\Leftrightarrow \qquad n = \frac{\log_{10}(10/7)}{\log_{10}(1.015)} \cong 24.$$

So $n \cong 24$ months $= 2$ years. It will require approximately 2 years to pay off the balance, assuming that payments of \$150 are made each month and the balance is not increased by any additional purchases.

d. Assuming that the person makes no additional purchases and pays \$150 each month, the person will have made 24 payments of \$150 each, for a total of \$3600 to pay off the initial balance of \$3000.

33. <u>Proof (by mathematical induction)</u>: : Let $f_1, f_2, f_3, \ldots$. be a sequence that satisfies the recurrence relation $f_k = f_{k-1} + 2^k$ for all integers $k \geq 2$, with initial condition $f_1 = 1$, and let the property $P(n)$ be the equation

$$f_n = 2^{n+1} - 3. \qquad \leftarrow P(n)$$

Show that $P(1)$ is true: The right-hand side of $P(1)$ is $2^{1+1} - 3 = 2^2 - 3 = 1$, which equals f_1, the left-hand side of $P(1)$. So $P(1)$ is true.

Show that for all integers $k \geq 1$, if $P(k)$ is true then $P(k+1)$ is true: Let k be any integer with $k \geq 1$, and suppose that

$$f_k = 2^{k+1} - 3. \qquad \leftarrow \begin{array}{l} P(k) \\ \text{inductive hypothesis} \end{array}$$

We must show that

$$f_{k+1} = 2^{(k+1)+1} - 3$$

or, equivalently,

$$f_{k+1} = 2^{k+2} - 3. \qquad \leftarrow P(k+1)$$

The left-hand side of $P(k+1)$ is

$$\begin{array}{rll} f_{k+1} & = & f_k + 2^{k+1} \qquad \text{by definition of } f_1, f_2, f_3, \ldots \\ & = & 2^{k+1} - 3 + 2^{k+1} \quad \text{by inductive hypothesis} \\ & = & 2 \cdot 2^{k+1} - 3 \\ & = & 2^{k+2} - 3 \qquad \text{by the laws of algebra,} \end{array}$$

which is the right-hand side of $P(k+1)$ [as to be shown].

36. Proof (by mathematical induction): Let $p_1, p_2, p_3, \ldots$ be a sequence that satisfies the recurrence relation $p_k = p_{k-1} + 2 \cdot 3^k$ for all integers $k \geq 2$, with initial condition $p_1 = 2$, and let the property $P(n)$ be the equation

$$p_n = 3^{n+1} - 7. \qquad \leftarrow P(n)$$

Show that $P(1)$ is true: The right-hand side of $P(1)$ is $3^{1+1} - 7 = 3^2 - 7 = 9 - 7 = 2$, which equals p_1, the left-hand side of $P(1)$. So $P(1)$ is true.

Show that for all integers $k \geq 1$, if $P(k)$ is true then $P(k+1)$ is true: Let k be any integer with $k \geq 1$, and suppose that

$$p_k = 3^{k+1} - 7. \qquad \leftarrow \begin{array}{l} P(k) \\ \text{inductive hypothesis} \end{array}$$

We must show that

$$p_{k+1} = 3^{(k+1)+1} - 7,$$

or, equivalently, that

$$p_{k+1} = 3^{k+2} - 7. \qquad \leftarrow P(k+1)$$

The left-hand side of $P(k+1)$ is

$$
\begin{array}{rll}
p_{k+1} & = & p_k + 2 \cdot 3^{k+1} \qquad \text{by definition of } p_1, p_2, p_3, \ldots \\
& = & (3^{k+1} - 7) + 2 \cdot 3^{k+1} \quad \text{by inductive hypothesis} \\
& = & 3^{k+1}(1 + 2) - 7 \\
& = & 3 \cdot 3^{k+1} - 7 \\
& = & 3^{k+2} - 7 \qquad \text{by the laws of algebra,}
\end{array}
$$

which is the right-hand side of $P(k+1)$ *[as to be shown]*.

42. Proof (by mathematical induction): Let $t_1, t_2, t_3, \ldots$ be a sequence that satisfies the recurrence relation $t_k = 2t_{k-1} + 2$ for all integers $k \geq 2$, with initial condition $t_1 = 2$, and let the property $P(n)$ be the equation $t_n = 2^{n+1} - 2$.

$$t_n = 2^{n+1} - 2. \qquad \leftarrow P(n)$$

Show that $P(1)$ is true: The right-hand side of $P(1)$ is $2^2 - 2 = 2$, which equals t_1, the left-hand side of $P(1)$. So $P(1)$ is true.

Show that for all integers $k \geq 1$, if $P(k)$ is true then $P(k+1)$ is true: Let k be any integer with $k \geq 1$, and suppose that

$$t_k = 2^{k+1} - 2. \qquad \leftarrow \begin{array}{l} P(k) \\ \text{inductive hypothesis} \end{array}$$

We must show that

$$t_{k+1} = 2^{(k+1)+1} - 2$$

or, equivalently, that

$$t_{k+1} = 2^{k+2} - 2. \qquad \leftarrow P(k+1)$$

The left-hand side of $P(k+1)$ is

$$
\begin{array}{rll}
t_{k+1} & = & 2t_k + 2 \qquad \text{by definition of } t_1, t_2, t_3, \ldots \\
& = & 2(2^{k+1} - 2) + 2 \quad \text{by inductive hypothesis} \\
& = & 2^{k+2} - 4 + 2 \\
& = & 2^{k+2} - 2 \qquad \text{by the laws of algebra,}
\end{array}
$$

which is the right-hand side of $P(k+1)$ *[as to be shown]*.

48. *a.*
$$w_1 = 1$$
$$w_2 = 2$$
$$w_3 = w_1 + 3 = 1 + 3$$
$$w_4 = w_2 + 4 = 2 + 4$$
$$w_5 = w_3 + 5 = 1 + 3 + 5$$
$$w_6 = w_4 + 6 = 2 + 4 + 6$$
$$w_7 = w_5 + 7 = 1 + 3 + 5 + 7$$

.

.

.

Guess: $w_n = \begin{cases} 1 + 3 + 5 + \cdots + n & \text{if } n \text{ is odd} \\ 2 + 4 + 6 + \cdots + n & \text{if } n \text{ is even} \end{cases}$

$= \begin{cases} \left(\dfrac{n+1}{2}\right)^2 & \text{if } n \text{ is odd} \\ 2\left(1 + 2 + 3 + \cdots + \dfrac{n}{2}\right) & \text{if } n \text{ is even} \end{cases}$ by exercise 5 of Section 5.2

$= \begin{cases} \left(\dfrac{n+1}{2}\right)^2 & \text{if } n \text{ is odd} \\ 2\left(\dfrac{\dfrac{n}{2}\left(\dfrac{n}{2}+1\right)}{2}\right) & \text{if } n \text{ is even} \end{cases}$ by Theorem 5.2.2

$= \begin{cases} \dfrac{(n+1)^2}{4} & \text{if } n \text{ is odd} \\ \dfrac{n(n+2)}{4} & \text{if } n \text{ is even} \end{cases}$ by the laws of algebra.

b. Proof (by strong mathematical induction): Let $w_1, w_2, w_3, \ldots.$ be a sequence that satisfies the recurrence relation $w_k = w_{k-2} + k$ for all integers $k \geq 3$, with initial conditions $w_1 = 1$ and $w_2 = 2$, and let the property $P(n)$ be the equation

$$w_n = \begin{cases} \dfrac{(n+1)^2}{4} & \text{if } n \text{ is odd} \\ \dfrac{n(n+2)}{4} & \text{if } n \text{ is even} \end{cases} \quad \text{for all integers } n \geq 1.$$

Show that $P(1)$ and $P(2)$ are true: For $n = 1$ and $n = 2$ the right-hand sides of $P(n)$ are

$$\frac{(1+1)^2}{4} = 1 \quad \text{and} \quad \frac{2(2+2)}{4} = 2,$$

which equal w_1 and w_2 respectively. So $P(1)$ and $P(2)$ are true.

Show that if $k \geq 2$ and $P(i)$ is true for all integers i from 1 through k, then $P(k+1)$ is true: Let k be any integer with $k \geq 2$, and suppose that

$$w_i = \begin{cases} \dfrac{(i+1)^2}{4} & \text{if } i \text{ is odd} \\ \dfrac{i(i+2)}{4} & \text{if } i \text{ is even} \end{cases} \quad \text{for all integers } i \text{ with } 1 \leq i < k. \quad \leftarrow \text{ inductive hypothesis}$$

We must show that

$$w_{k+1} = \begin{cases} \dfrac{(k+2)^2}{4} & \text{if } k+1 \text{ is odd} \\ \dfrac{(k+1)(k+3)}{4} & \text{if } k+1 \text{ is even} \end{cases}$$

But

$$w_{k+1} \;=\; w_{k-1} + (k+1) \qquad \text{by definition of } w_1, w_2, w_3, \ldots$$

$$= \begin{cases} \dfrac{((k-1)+1)^2}{4} + (k+1) & \text{if } k-1 \text{ is odd} \\[2mm] \dfrac{(k-1)[(k-1)+2]}{4} + (k+1) & \text{if } k-1 \text{ is even} \end{cases} \qquad \text{by inductive hypothesis}$$

$$= \begin{cases} \dfrac{k^2}{4} + \dfrac{4(k+1)}{4} & \text{if } k+1 \text{ is odd} \\[2mm] \dfrac{(k-1)(k+1)}{4} + \dfrac{4(k+1)}{4} & \text{if } k+1 \text{ is even} \end{cases} \qquad \begin{array}{l}\text{because } k-1 \text{ and } k+1 \text{ have}\\ \text{the same parity}\end{array}$$

$$= \begin{cases} \dfrac{k^2 + 4k + 4}{4} & \text{if } k+1 \text{ is odd} \\[2mm] \dfrac{k^2 + 4k + 3}{4} & \text{if } k+1 \text{ is even} \end{cases}$$

$$= \begin{cases} \dfrac{(k+2)^2}{4} & \text{if } k+1 \text{ is odd} \\[2mm] \dfrac{(k+1)(k+3)}{4} & \text{if } k+1 \text{ is even} \end{cases} \qquad \text{by the laws of algebra.}$$

[This is what was to be shown.]

51. The sequence does not satisfy the formula. By definition of $a_1, a_2, a_3, \ldots$, $a_1 = 0$, $a_2 = (a_1+1)^2 = 1^2 = 1$, $a_3 = (a_2+1)^2 = (1+1)^2 = 4$, $a_4 = (a_3+1)^2 = (4+1)^2 = 25$. But according to the formula $a_4 = (4-1)^2 = 9 \neq 25$.

54. *a.*

$$Y_1 = E + c + mY_0$$

$$Y_2 = E + c + mY_1 = E + c + m(E + c + mY_0) = (E + c) + m(E + c) + m^2 Y_0$$

$$Y_3 = E + c + mY_2 = E + c + m((E+c) + m(E+c) + m^2 Y_0) = (E+c) + m(E+c) + m^2(E+c) + m^3 Y_0$$

$$Y_4 = E + c + mY_3 = E + c + m((E+c) + m(E+c) + m^2(E+c) + m^3 Y_0)$$

$$= (E+c) + m(E+c) + m^2(E+c) + m^3(E+c) + m^4 Y_0$$

.

.

.

Guess: $Y_n = (E+c) + m(E+c) + m^2(E+c) + \cdots + m^{n-1}(E+c) + m^n Y_0$

$$= (E+c)[1 + m + m^2 + \cdots + m^{n-1}] + m^n Y_0$$

$$= (E+c)\left(\dfrac{m^n - 1}{m - 1}\right) + m^n Y_0, \text{ for all integers } n \geq 1.$$

b. Suppose $0 < m < 1$. Then

$$
\begin{aligned}
\lim_{n\to\infty} Y_n &= \lim_{n\to\infty}\left((E+c)\left(\frac{m^n-1}{m-1}\right)+m^n Y_0\right) \\
&= (E+c)\left(\frac{\lim_{n\to\infty} m^n-1}{m-1}\right)+\lim_{n\to\infty} m^n Y_0 \\
&= (E+c)\left(\frac{0-1}{m-1}\right)+0\cdot Y_0 \qquad\qquad \text{because when } 0 < m < 1, \\
&\qquad\qquad\qquad\qquad\qquad\qquad\qquad \text{then } \lim_{n\to\infty} m^n = 0 \\
&= \frac{E+c}{1-m}.
\end{aligned}
$$

Section 5.8

3. b.
$$
\left\{\begin{array}{l} a_0 = C\cdot 2^0 + D = C + D = 0 \\ a_1 = C\cdot 2^1 + D = 2C + D = 2 \end{array}\right\} \Leftrightarrow \left\{\begin{array}{l} D = -C \\ 2C + (-C) = 2 \end{array}\right\} \Leftrightarrow \left\{\begin{array}{l} D = -C \\ C = 2 \end{array}\right\} \Leftrightarrow \left\{\begin{array}{l} C = 2 \\ D = -2 \end{array}\right\}
$$
$$a_2 = C\cdot 2^2 + D = 2\cdot 2^2 + (-2) = 6$$

6. <u>Proof:</u> Given that $b_n = C\cdot 3^n + D(-2)^n$, then for any choice of C and D and integer $k \geq 2$,

$$b_k = C\cdot 3^k + D\cdot(-2)^k, \quad b_{k-1} = C\cdot 3^{k-1} + D\cdot(-2)^{k-1}, \quad \text{and} \quad b_{k-2} = C\cdot 3^{k-2} + D\cdot(-2)^{k-2}.$$

Hence,

$$
\begin{aligned}
b_{k-1} + 6b_{k-2} &= \left(C\cdot 3^{k-1} + D(-2)^{k-1}\right) + 6\left(C\cdot 3^{k-2} + D\cdot(-2)^{k-2}\right) \\
&= C\cdot\left(3^{k-1} + 6\cdot 3^{k-2}\right) + D\cdot\left((-2)^{k-1} + 6\cdot(-2)^{k-2}\right) \\
&= C\cdot 3^{k-2}(3+6) + D\cdot(-2)^{k-2}(-2+6) \\
&= C\cdot 3^{k-2}\cdot 3^2 + D\cdot(-2)^{k-2}2^2 \\
&= C\cdot 3^k + D\cdot(-2)^k \\
&= b_k.
\end{aligned}
$$

9. a. If for all integers $k \geq 2$, $t^k = 7t^{k-1} - 10t^{k-2}$ and $t \neq 0$, then $t^2 = 7t - 10$ and so $t^2 - 7t + 10 = 0$. But $t^2 - 7t + 10 = (t-2)(t-5)$. Thus $t = 2$ or $t = 5$.

b. It follows from part (a) and the distinct roots theorem that for some constants C and D, the terms of the sequence $b_0, b_1, b_2, \ldots$ satisfy the equation

$$b_n = C\cdot 2^n + D\cdot 5^n \quad \text{for all integers } n \geq 0.$$

Since $b_0 = 2$ and $b_1 = 2$, then
$$
\left\{\begin{array}{l} b_0 = C\cdot 2^0 + D\cdot 5^0 = C + D = 2 \\ b_1 = C\cdot 2^1 + D\cdot 5^1 = 2C + 5D = 2 \end{array}\right\} \Leftrightarrow \left\{\begin{array}{l} D = 2 - C \\ 2C + 5(2-C) = 2 \end{array}\right\}
$$
$$
\Leftrightarrow \left\{\begin{array}{l} D = 2 - C \\ C = 8/3 \end{array}\right\} \qquad\qquad \Leftrightarrow \left\{\begin{array}{l} D = 2 - (8/3) = -(2/3) \\ C = 8/3 \end{array}\right\}
$$
Thus $b_n = \dfrac{8}{3}\cdot 2^n - \dfrac{2}{3}\cdot 5^n$ for all integers $n \geq 0$.

12. The characteristic equation is $t^2 - 9 = 0$. Since $t^2 - 9 = (t-3)(t+3)$, the roots are $t = 3$ and $t = -3$. By the distinct roots theorem, there exist constants C and D such that

$$e_n = C\cdot 3^n + D\cdot(-3)^n \quad \text{for all integers } n \geq 0.$$

Since $e_0 = 0$ and $e_1 = 2$, then

$$\left\{\begin{array}{l} e_0 = C\cdot 3^0 + D\cdot(-3)^0 = C + D = 0 \\ e_1 = C\cdot 3^1 + D\cdot(-3)^1 = 3C - 3D = 2 \end{array}\right\} \Leftrightarrow \left\{\begin{array}{l} D = -C \\ 3C - 3(-C) = 2 \end{array}\right\} \Leftrightarrow \left\{\begin{array}{l} D = -1/3 \\ C = 1/3 \end{array}\right\}$$

Thus $e_n = \dfrac{1}{3}\cdot 3^n - \dfrac{1}{3}(-3)^n = 3^{n-1} + (-3)^{n-1} = 3^{n-1}(1 + (-1)^{n-1}) = \left\{\begin{array}{ll} 2\cdot 3^{n-1} & \text{if } n \text{ is odd} \\ 0 & \text{if } n \text{ is even} \end{array}\right.$

for all integers $n \geq 0$.

15. The characteristic equation is $t^2 - 6t + 9 = 0$. Since $t^2 - 6t + 9 = (t-3)^2$, there is only one root, $t = 3$. By the single root theorem, there exist constants C and D such that

$$t_n = C\cdot 3^n + D\cdot n\cdot 3^n \quad \text{for all integers } n \geq 0.$$

Since $t_0 = 1$ and $t_1 = 3$, then

$$\left\{\begin{array}{l} t_0 = C\cdot 3^0 + D\cdot 0\cdot 3^0 = C = 1 \\ t_1 = C\cdot 3^1 + D\cdot 1\cdot 3^1 = 3C + 3D = 3 \end{array}\right\} \Leftrightarrow \left\{\begin{array}{l} C = 1 \\ C + D = 1 \end{array}\right\} \Leftrightarrow \left\{\begin{array}{l} C = 1 \\ D = 0 \end{array}\right\}$$

Thus $t_n = 1\cdot 3^n + 0\cdot n\cdot 3^n = 3^n$ for all integers $n \geq 0$.

18. <u>Proof</u>: Suppose that $s_0, s_1, s_2, \ldots$ and $t_0, t_1, t_2, \ldots$ are sequences such that for all integers $k \geq 2$,

$$s_k = 5s_{k-1} - 4s_{k-2} \quad \text{and} \quad t_k = 5t_{k-1} - 4t_{k-2}.$$

Then for all integers $k \geq 2$,

$$\begin{aligned} 5(2s_{k-1} + 3t_{k-1}) - 4(2s_{k-2} + 3t_{k-2}) &= (5\cdot 2s_{k-1} - 4\cdot 2s_{k-2}) + (5\cdot 3t_{k-1} - 4\cdot 3t_{k-2}) \\ &= 2(5s_{k-1} - 4s_{k-2}) + 3(5t_{k-1} - 4t_{k-2}) \\ &= 2s_k + 3t_k. \end{aligned}$$

[This is what was to be shown.]

21. Let $a_0, a_1, a_2, \ldots$ be any sequence that satisfies the recurrence relation $a_k = Aa_{k-1} + Ba_{k-2}$ for some real numbers A and B with $B \neq 0$ and for all integers $k \geq 2$. Furthermore, suppose that the equation $t^2 - At - B = 0$ has a single real root r. First note that $r \neq 0$ because otherwise we would have $0^2 - A\cdot 0 - B = 0$, which would imply that $B = 0$ and contradict the hypothesis. Second, note that the following system of equations with unknowns C and D has a unique solution.

$$a_0 = Cr^0 + 0\cdot Dr^0 = 1\cdot C + 0\cdot D$$
$$a_1 = Cr^1 + 1\cdot Dr^1 = C\cdot r + D\cdot r$$

One way to reach this conclusion is to observe that the determinant of the system is $1\cdot r - r\cdot 0 = r \neq 0$. Another way to reach the conclusion is to write the system as

$$a_0 = C$$
$$a_1 = Cr + Dr$$

and let $C = a_0$ and $D = (a_1 - a_0 r)/r$. It is clear by substitution that these values of C and D satisfy the system. Conversely, if any numbers C and D satisfy the system, then $C = a_0$, and substituting C into the second equation and solving for D yields $D = (a_1 - Cr)/r$.

<u>Proof of the exercise statement by strong mathematical induction</u>: Let $a_0, a_1, a_2, \ldots$ be any sequence that satisfies the recurrence relation $a_k = Aa_{k-1} + Ba_{k-2}$ for some real numbers A and B with $B \neq 0$ and for all integers $k \geq 2$. Furthermore, suppose that the equation $t^2 - At - B = 0$ has a single real root r. Let the property $P(n)$ be the equation

$$a_n = Cr^n + nDr^n \qquad \leftarrow P(n)$$

where C and D are the unique real numbers such that $a_0 = Cr^0 + 0\cdot Dr^0$ and $a_1 = Cr^1 + 1\cdot Dr^1$.

***Show that $P(0)$ and $P(1)$ are true:* :** The fact $P(0)$ and $P(1)$ are true is automatic because C and D are exactly those numbers for which $a_0 = Cr^0 + 0 \cdot Dr^0$ and $a_1 = C \cdot r^1 + 1 \cdot Dr^1$.

Show that if $k \geq 1$ and $P(i)$ is true for all integers i from 0 through k, then $P(k+1)$ is true: Let k be any integer with $k \geq 1$ and suppose that

$$a_i = Cr^i + iDr^i \quad \text{for all integers } i \text{ with } 1 \leq i \leq k. \quad \leftarrow \text{inductive hypothesis}$$

We must show that

$$a_{k+1} = Cr^{k+1} + (k+1)Dr^{k+1}.$$

Now by the inductive hypothesis,

$$a_k = Cr^k + kDr^k \quad \text{and} \quad a_{k-1} = Cr^{k-1} + (k-1)Dr^{k-1}.$$

So

$$
\begin{aligned}
a_{k+1} &= Aa_k + Ba_{k-1} & \text{by definition of } a_0, a_1, a_2, \dots \\
&= A(Cr^k + kDr^k) + B(Cr^{k-1} + (k-1)Dr^{k-1}) & \\
& & \text{by inductive hypothesis} \\
&= C(Ar^k + Br^{k-1}) + D(Akr^k + B(k-1)r^{k-1}) & \\
& & \text{by algebra} \\
&= Cr^{k+1} + Dkr^{k+1} & \text{by Lemma 5.8.4.}
\end{aligned}
$$

[This is what was to be shown.]

24. *a.* If $\dfrac{\phi}{1} = \dfrac{1}{\phi - 1}$, then $\phi(\phi - 1) = 1$, or, equivalently, $\phi^2 - \phi - 1 = 0$ and so ϕ satisfies the equation $t^2 - t - 1 = 0$.

b. By the quadratic formula, the solutions to $t^2 - t - 1 = 0$ are

$$t = \frac{1 \pm \sqrt{1 + 4}}{2} = \begin{cases} \dfrac{1 + \sqrt{5}}{2} \\[2mm] \dfrac{1 - \sqrt{5}}{2}. \end{cases}$$

Let

$$\phi_1 = \frac{1 + \sqrt{5}}{2} \quad \text{and} \quad \phi_2 = \frac{1 - \sqrt{5}}{2}.$$

c. $F_n = \dfrac{1}{\sqrt{5}} \cdot \phi_1^{n+1} - \dfrac{1}{\sqrt{5}} \cdot \phi_2^{n+1} = \dfrac{1}{\sqrt{5}}(\phi_1^{n+1} - \phi_2^{n+1})$

This equation is an alternative way to write equation (5.8.8).

Section 5.9

3. *b.* (1) MI is in the MIU-system by I.

(2) MII is in the MIU-system by (1) and $II(b)$.

(3) $MIIII$ is in the MIU-system by (2) and $II(b)$.

(4) $MIIIIIIII$ is in the MIU-system by (3) and $II(b)$.

(5) $MUIIIII$ is in the MIU-system by (4) and $II(c)$.

(6) $MUIIU$ is in the MIU-system by (5) and $II(c)$.

6. <u>Proof (by structural induction)</u>: Let the property be the following sentence: The string begins with an a.

   Show that each object in the BASE for S satisfies the property: The only object in the base is a, and the string a begins with an a.

   Show that for each rule in the RECURSION for S, if the rule is applied to objects in S that satisfy the property, then the objects defined by the rule also satisfy the property:

   The recursion for S consists of two rules, denoted II(a) and II(b).

   In case rule II(a) is applied to a string s in S that begins with a a, the result is the string sa, which begins with the same character as s, namely a.

   Similarly, in case rule II(b) is applied to a string s that begins with a a, the result is the string sb, which also begins with an a.

   Thus, when each rule in the RECURSION is applied to strings in S that begin with an a, the results are also strings that begin with an a.

   Because no objects other than those obtained through the BASE and RECURSION conditions are contained in S, every string in S begins with an a.

9. <u>Proof (by structural induction)</u>: Let the property be the following sentence: The string represents an odd integer.

   Show that each object in the BASE for S satisfies the property: The objects in the base are 1, 3, 5, 7, and 9. All of these strings represent odd integers.

   Show that for each rule in the RECURSION for S, if the rule is applied to objects in S that satisfy the property, then the objects defined by the rule also satisfy the property:

   The recursion for S consists of five rules, denoted II(a)–II(e).

   Suppose s and t are strings in S that represent odd integers.

   Then the right-most character for each of s and t is 1, 3, 5, 7, or 9.

   In case rule II(a) is applied to s and t, the result is the string st, which has the same right-most character as t. So st represents an odd integer.

   In case rules II(b)–II(e) are applied to s, the results are $2s$, $4s$, $6s$, or $8s$. All of these strings have the same right-most character as s, and, therefore, they all represent odd integers.

   Thus when each rule in the RECURSION is applied to strings in S that represent odd integers, the result is also a string that represents an odd integer.

   Because no objects other than those obtained through the BASE and RECURSION conditions are contained in S, all the strings in S represent odd integers.

12. The string MU is not in the system because the number of I's in MU is 0, which is divisible by 3, and for all strings in the MIU-system, the number of I's in the string is not divisible by 3.

   <u>Proof (by structural induction)</u>: Let the property be the following sentence: The number of I's in the string is not divisible by 3.

   Show that each object in the BASE for the MIU-system satisfies the property: The only object in the base is MI, which has one I, and the number 1 is not divisible by 3.

   Show that for each rule in the RECURSION for the MIU-system , if the rule is applied to objects in the system that satisfy the property, then the objects defined by the rule also satisfy the property:

The recursion for the MIU-system consists of four rules, denoted II(a)–(d). Let s be a string, let n be the number of I's in s, and suppose $3 \nmid n$. Consider the effect of acting upon s by each recursion rule in turn.

In case rule II(a) is applied to s, s has the form xI, where x is a string. The result is the string xIU. This string has the same number of I's as xI, namely n, and n is not divisible by 3.

In case rule II(b) is applied to s, s has the form Mx, where x is a string. The result is the string Mxx. This string has twice the number of I's as Mx. Because n is the number of I's in Mx and $3 \nmid n$, we have $n = 3k + 1$ or $n = 3k + 2$ for some integer k. In case $n = 3k + 1$, the number of I's in Mxx is $2(3k + 1) = 3(2k) + 2$, which is not divisible by 3. In case $n = 3k + 2$, the number of I's in Mxx is $2(3k + 2) = 6k + 4 = 3(2k + 1) + 1$, which is not divisible by 3 either.

In case rule II(c) is applied to s, s has the form $xIIIy$, where x and y are strings. The result is the string xUy. This string has three fewer I's than the number of I's in s. Because n is the number of I's in $xIIIy$ and $3 \nmid n$, we have that $3 \nmid (n - 3)$ either *[for if $3 \mid (n - 3)$ then $n - 3 = 3k$, for some integer k. Hence $n = 3k + 3 = 3(k + 1)$, and so n would be divisible by 3, which it is not]*. Thus the number of I's in xUy is not divisible by 3.

In case rule II(d) is applied to s, s has the form $xUUy$, where x and y are strings. The result is the string xUy. This string has the same number of I's as $xUUy$, namely n, and n is not divisible by 3.

By the restriction for the MIU-system, no strings other than those derived from the base and the recursion are in the system. Therefore, for all strings in the MIU-system, the number of I's in the string is not divisible by 3.

18. Let S be the set of all strings of a's and $b's$ that contain exactly one a. The following is a recursive definition of S.

    I. BASE: $a \in S$

    II. RECURSION: If $s \in S$, then

    a. $bs \in S$ b. $sb \in S$

    III. RESTRICTION: There are no elements of S other than those obtained from I and II.

21. *b.* $A(2,1) = A(1, A(2,0)) = A(1, A(1,1)) = A(1,3)$ *[by part (a)]* $= A(0, A(1,2)) = A(0,4)$ *[by Example 5.9.9]* $= 4 + 1 = 5$

Review Guide: Chapter 5

Sequences and Summations

- What is a method that can sometimes find an explicit formula for a sequence whose first few terms are given (provided a nice explicit formula exists!)? *(p. 229)*
- What is the expanded form for a sum that is given in summation notation? *(p. 231)*
- What is the summation notation for a sum that is given in expanded form? *(p. 231)*
- How do you evaluate $a_1 + a_2 + a_3 + \cdots a_n$ when n is small? *(p. 232)*
- What does it mean to "separate off the final term of a summation"? *(p. 232)*
- What is the product notation? *(p. 233)*
- What are some properties of summations and products? *(p. 234)*
- How do you transform a summation by making a change of variable? *(p. 235)*
- What is factorial notation? *(p. 237)*
- What is the n choose r notation? *(p. 238)*
- What is an algorithm for converting from base 10 to base 2? *(p. 241)*
- What does it mean for a sum to be written in closed form? *(p. 251)*

Mathematical Induction

- What do you show in the basis step and what do you show in the inductive step when you use (ordinary) mathematical induction to prove that a property involving an integer n is true for all integers greater than or equal to some initial integer? *(p. 247)*
- What is the inductive hypothesis in a proof by (ordinary) mathematical induction? *(p. 247)*
- Are you able to use (ordinary) mathematical induction to construct proofs involving various kinds of statements such as formulas, divisibility properties, inequalities, and other situations? *(pp. 247, 249, 253, 259, 263, 265,)*
- Are you able to apply the formula for the sum of the first n positive integers? *(p. 251)*
- Are you able to apply the formula for the sum of the successive powers of a number, starting with the zeroth power? *(p. 255)*

Strong Mathematical Induction and The Well-Ordering Principle for the Integers

- What do you show in the basis step and what do you show in the inductive step when you use strong mathematical induction to prove that a property involving an integer n is true for all integers greater than or equal to some initial integer? *(p. 268)*
- What is the inductive hypothesis in a proof by strong mathematical induction? *(p. 268)*
- Are you able to use strong mathematical induction to construct proofs of various statements? *(pp. 269-274)*
- What is the well-ordering principle for the integers? *(p. 275)*
- Are you able to use the well-ordering principle for the integers to prove statements, such as the existence part of the quotient-remainder theorem? *(p. 276)*
- How are ordinary mathematical induction, strong mathematical induction, and the well-ordering principle for the integers related? *(p. 275 and exercises 31 and 32 on p. 279)*

Algorithm Correctness

- What are the pre-condition and the post-condition for an algorithm? *(p. 280)*
- What does it mean for a loop to be correct with respect to its pre- and post-conditions? *(p. 281)*

- What is a loop invariant? *(p. 282)*
- How do you use the loop invariant theorem to prove that a loop is correct with respect to its pre- and post-conditions? *(pp. 283-288)*

Recursion

- What is an explicit formula for a sequence? *(p. 290)*
- What does it mean to define a sequence recursively? *(p. 290)*
- What is a recurrence relation with initial conditions? *(p. 290)*
- How do you compute terms of a recursively defined sequence? *(p. 290)*
- Can different sequences satisfy the same recurrence relation? *(p. 291)*
- What is the "recursive paradigm"? *(p. 293)*
- How do you develop a recurrence relation for the tower of Hanoi sequence? *(p. 294)*
- How do you develop a recurrence relations for the Fibonacci sequence? *(p. 297)*
- How do you develop recurrence relations for sequences that involve compound interest? *(pp. 298-299)*
- How do you mathematical induction to prove properties of summations? *(p. 301)*

Solving Recurrence Relations

- What is the method of iteration for solving a recurrence relation? *(p. 305)*
- What is an arithmetic sequence? *(p. 307)*
- What is a geometric sequence? *(p. 308)*
- How do you use the formula for the sum of the first n integers and the formula for the sum of the first n powers of a real number r to simplify the answers you obtain when you solve recurrence relations? *(pp. 309-310)*
- How is mathematical induction used to check that the solution to a recurrence relation is correct? *(p. 312-314)*
- What is a second-order linear homogeneous recurrence relation with constant coefficients? *(p. 317)*
- What is the characteristic equation for a second-order linear homogeneous recurrence relation with constant coefficients? *(p. 319)*
- What is the distinct-roots theorem? If the characteristic equation of a relation has two distinct roots, how do you solve the relation? *(p. 321)*
- What is the single-root theorem? If the characteristic equation of a relation has a single root, how do you solve the relation? *(p. 325)*

General Recursive Definitions

- When a set is defined recursively, what are the three parts of the definition? *(p. 328)*
- Given a recursive definition for a set, how can you tell that a given element is in the set? *(p. 328-329)*
- What is structural induction? *(p. 331)*
- Given a recursive definition for a set, is there a way to tell that a given element is not in the set? *(solution for exercise 14a on p. A-49)*
- What is a recursive function? *(p. 332)*

Formats for Proving Formulas by Mathematical Induction

When using mathematical induction to prove a formula, students are sometimes tempted to present their proofs in a way that assumes what is to be proved. There are several formats you can use, besides the one shown most frequently in the textbook, to avoid this fallacy. A crucial point is this:

> If you are hoping to prove that an equation is true but you haven't yet done so, either preface it with the words "We must show that" or put a question mark above the equal sign.

Format 1 (the format used most often in the textbook for the inductive step): Start with the left-hand side (LHS) of the equation to be proved and successively transform it using definitions, known facts from basic algebra, and (for the inductive step) the inductive hypothesis until you obtain the right-hand side (RHS) of the equation.

Format 2 (the format used most often in the textbook for the basis step): Transform the LHS and the RHS of the equation to be proved *independently*, one after the other, until both sides are shown to equal the same expression. Because two quantities equal to the same quantity are equal to each other, you can conclude that the two sides of the equation are equal to each other.

Format 3: This format is just like Format 2 except that the computations are done in parallel. But in order to avoid the fallacy of assuming what is to be proved, do NOT put an equal sign between the two sides of the equation until the very last step. Separate the two sides of the equation with a vertical line.

Format 4: This format is just like Format 3 except that the two sides of the equation are separated by an equal sign with a question mark on top: $\overset{?}{=}$

Format 5: Start by writing something like "We must show that" and the equation you want to prove true. In successive steps, indicate that this equation is true if, and only if, ($\Leftrightarrow$) various other equations are true. But be sure that both the directions of your "if and only if" claims are correct. In other words, be sure that the $\Leftarrow$ direction is just as true as the $\Rightarrow$ direction. If you finally get down to an equation that is known to be true, then because each subsequent equation is true *if, and only if*, the previous equation is true, you will have shown that the original equation is true.

Example: Let the property $P(n)$ be the equation

$$\boxed{1 + 3 + 5 + \cdots + (2n - 1) = n^2} \leftarrow P(n).$$

Proof that $P(1)$ is true:

Solution (Format 2):
When $n = 1$, the LHS of $P(1)$ equals 1, and the RHS equals 1^2 which also equals 1. So $P(1)$ is true.

Proof that for all integers $k \geq 1$, if $P(k)$ is true then $P(k + 1)$ is true:

Solution (Format 2):
Suppose that k is any integer with $k \geq 1$ such that $1 + 3 + 5 + \cdots + (2k - 1) = k^2$. *[This is the inductive hypothesis, $P(k)$.]* We must show that $P(k + 1)$ is true, where $P(k + 1)$ is the equation $1 + 3 + 5 + \cdots + (2k + 1) = (k + 1)^2$.

Now the LHS of $P(k + 1)$ is

$$1 + 3 + 5 + \cdots + (2k + 1) = 1 + 3 + 5 + \cdots + (2k - 1) + (2k + 1)$$
$$\text{by making the next-to-last term explicit}$$
$$= k^2 + (2k + 1) \quad \text{by inductive hypothesis.}$$

And the RHS of $P(k + 1)$ is

$$(k + 1)^2 = k^2 + 2k + 1 \quad \text{by basic algebra.}$$

So the left-hand and right-hand sides of $P(k+1)$ equal the same quantity, and thus and thus $P(k+1)$ is true *[as was to be shown]*.

Solution (Format 3):

Suppose that k is any integer with $k \geq 1$ such that $1 + 3 + 5 + \cdots + (2k - 1) = k^2$. *[This is the inductive hypothesis, $P(k)$.]* We must show that $P(k + 1)$ is true, where $P(k + 1)$ is the equation $1 + 3 + 5 + \cdots + (2k + 1) = (k + 1)^2$.

Consider the left-hand and right-hand sides of $P(k + 1)$:

$$
\begin{array}{l|l}
1 + 3 + 5 + \cdots + (2k + 1) & (k + 1)^2 \\
= 1 + 3 + 5 + \cdots + (2k - 1) + (2k + 1) & \\
\quad \text{by making the next-to-last term explicit} & \\
= k^2 + (2k + 1) & \\
\quad \text{by inductive hypothesis} & \\
= k^2 + 2k + 1 & = k^2 + 2k + 1 \\
\quad \text{by basic algebra} & \quad \text{by basic algebra}
\end{array}
$$

So the left-hand and right-hand sides of $P(k+1)$ equal the same quantity, and thus and thus $P(k+1)$ is true *[as was to be shown]*.

Solution (Format 4):

Suppose that k is any integer with $k \geq 1$ such that $1 + 3 + 5 + \cdots + (2k - 1) = k^2$. *[This is the inductive hypothesis, $P(k)$.]* We must show that $P(k + 1)$ is true, where $P(k + 1)$ is the equation $1 + 3 + 5 + \cdots + (2k + 1) = (k + 1)^2$.

Consider the left-hand and right-hand sides of $P(k + 1)$:

$$
\begin{array}{ll}
1 + 3 + 5 + \cdots + (2k + 1) & \overset{?}{=} \quad (k + 1)^2 \\
1 + 3 + 5 + \cdots + (2k - 1) + (2k + 1) & \overset{?}{=} \quad k^2 + 2k + 1 \\
\quad \text{by making the next-to-last term explicit} & \qquad \text{by basic algebra} \\
k^2 + (2k + 1) & \overset{?}{=} \quad k^2 + 2k + 1 \\
\quad \text{by inductive hypothesis} & \\
k^2 + 2k + 1 & = \quad k^2 + 2k + 1 \\
\quad \text{by basic algebra} &
\end{array}
$$

So the left-hand and right-hand sides of $P(k + 1)$ equal the same quantity, and thus $P(k + 1)$ is true *[as was to be shown]*.

Solution (Format 5):

Suppose that k is any integer with $k \geq 1$ such that $1 + 3 + 5 + \cdots + (2k - 1) = k^2$. *[This is the inductive hypothesis, $P(k)$.]* We must show that $P(k + 1)$ is true, where $P(k + 1)$ is the equation $1 + 3 + 5 + \cdots + (2k + 1) = (k + 1)^2$.

But $P(k + 1)$ is true if, and only if, $(\Leftrightarrow)$

$$
\begin{array}{rrll}
& 1 + 3 + 5 + \cdots + (2k - 1) + (2k + 1) & = & (k + 1)^2 \quad \text{by making the next-to-last term explicit} \\
\Leftrightarrow & k^2 + (2k + 1) & = & (k + 1)^2 \quad \text{by inductive hypothesis} \\
\Leftrightarrow & k^2 + 2k + 1 & = & (k + 1)^2
\end{array}
$$

which is true by basic algebra. Thus $P(k + 1)$ is true *[as was to be shown]*.

Chapter 6: Set Theory

The first section of this chapter introduces additional terminology for sets and the concept of an element argument to prove that one set is a subset of another. The aim of this section is to provide a experience with a variety of types of sets and a basis for deriving the set properties discussed in the remainder of the chapter. The second and third sections show how to prove and disprove various proposed set properties of union, intersection, set difference and (general) complement using element arguments, algebraic arguments, and counterexamples. Section 5.4 introduces the concept of Boolean algebra, which generalizes both the algebra of sets with the operations of union and intersection and the properties of a set of statements with the operations of *or* and *and*. The section goes on to discuss Russell's paradox and shows that reasoning similar to Russell's can be used to prove an important property of computer algorithms.

Section 6.1

3. *c.* Yes. Every element in T is in S because every integer that is divisible by 6 is also divisible by 3. To see why this is so, suppose n is any integer that is divisible by 6. Then $n = 6m$ for some integer m. Since $6m = 3(2m)$ and since $2m$ is an integer (being a product of integers), it follows that $n = 3 \cdot$ (some integer), and, hence, that n is divisible by 3.

6. *a.* $A \nsubseteq B$ because $2 \in A$ (because $2 = 5 \cdot 0 + 2$) but $2 \notin B$ (because if $2 = 10b - 3$ for some integer b, then $10b = 5$, so $b = 1/2$, which is not an integer).

   b. $B \subseteq A$

   Proof:

   Suppose y is a particular but arbitrarily chosen element of B.

   [We must show that y is in A. By definition of A, this means that we must show that $y = 5 \cdot$(some integer) + 2.]

   By definition of B, $y = 10b - 3$ for some integer b.

   [Scratch work: Is there an integer, say a, such that $y = 5a + 2$? If so, then $5a + 2 = 10b - 3$, which implies that $5a = 10b - 5$, or, equivalently, that $a = 2b - 1$. So give this value to a and see if it works.]

   Let $a = 2b - 1$. Then a is an integer and $5a + 2 = 5(2b - 1) + 2 = 10b - 5 + 2 = 10b - 3 = y$. Thus y is in A *[as was to be shown]*.

   c. $B = C$

   Proof:

   Part 1, Proof That $B \subseteq C$:

   Suppose y is a particular but arbitrarily chosen element of B.

   [We must show that y is in C. By definition of C, this means that we must show that $y = 10 \cdot$(some integer) + 7.]

   By definition of B, $y = 10b - 3$ for some integer b.

   [Scratch work: Is there an integer, say c, such that $y = 10c + 7$? If so, then $10c + 7 = 10b - 3$, which implies that $10c = 10b - 10$, or, equivalently, that $c = b - 1$. So give this value to c and see if it works.]

   Let $c = b - 1$. Then c is an integer and $10c + 7 = 10(b - 1) + 7 = 10b - 10 + 7 = 10b - 3 = y$. Thus y is in C *[as was to be shown]*.

Part 2, Proof That $C \subseteq B$:

Suppose z is a particular but arbitrarily chosen element of C.

[We must show that z is in B. By definition of B, this means that we must show that $z = 10\cdot(\text{some integer}) - 3$.]

By definition of C, $z = 10c + 7$ for some integer c.

[Scratch work: Is there an integer, say b, such that $z = 10b - 3$? If so, then $10b - 3 = 10c + 7$, which implies that $10b = 10c + 10$, or, equivalently, that $b = c + 1$. So give this value to b and see if it works.]

9. *b.* $x \notin A$ or $x \notin B$ *c.* $x \notin A$ or $x \in B$

12.

     a. $A \cup B = \{x \in \mathbf{R} \mid -3 \leq x < 2\}$ *b.* $A \cap B = \{x \in \mathbf{R} \mid -1 < x \leq 0\}$
     c. $A^c = \{x \in \mathbf{R} \mid x < -3 \text{ or } x > 0\}$ *d.* $A \cup C = \{x \in \mathbf{R} \mid -3 \leq x \leq 0 \text{ or } 6 < x \leq 8\}$
     e. $A \cap C = \emptyset$ *f.* $B^c = \{x \in \mathbf{R} \mid x \leq -1 \text{ or } x \geq 2\}$
     g. $A^c \cap B^c = \{x \in \mathbf{R} \mid x < -3 \text{ or } x \geq 2\}$ *h.* $A^c \cup B^c = \{x \in \mathbf{R} \mid x \leq -1 \text{ or } x > 0\}$
     i. $(A \cap B)^c = \{x \in \mathbf{R} \mid x \leq -1 \text{ or } x > 0\}$ *j.* $(A \cup B)^c = \{x \in \mathbf{R} \mid x < -3 \text{ or } x \geq 2\}$

     Note that $(A \cap B)^c = A^c \cup B^c$ and that $(A \cup B)^c = A^c \cap B^c$.

15. *b.*

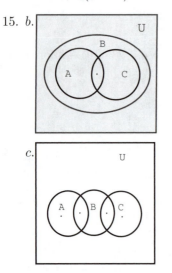

    c.

18. *c.* Yes, because $\{\emptyset\}$ is the set that contains the one element $\emptyset$.

     d. No, because $\emptyset$ has no elements and thus it cannot contain the element $\emptyset$.

27. *b.* Yes. Every element in $\{p, q, u, v, w, x, y, z\}$ is in one of the sets of the partition and no element is in more than one set of the partition.

     c. No. The number 4 is in both sets $\{5, 4\}$ and $\{1, 3, 4\}$.

     e. Yes. Every element in $\{1, 2, 3, 4, 5, 6, 7, 8\}$ is in one of the sets of the partition and no element is in more than one set of the partition.

30. Yes. By the quotient-remainder theorem, every integer can be represented in exactly one of the following forms: $4k$ or $4k+1$ or $4k+2$ or $4k+3$ for some integer k. Thus $\mathbf{Z} = A_0 \cup A_1 \cup A_2 \cup A_3$,

     $A_0 \cap A_1 = \emptyset$, $A_0 \cap A_2 = \emptyset$, $A_0 \cap A_3 = \emptyset$, $A_1 \cap A_2 = \emptyset$, $A_1 \cap A_3 = \emptyset$, and $A_2 \cap A_3 = \emptyset$.

33. *a.* $\mathscr{P}(\emptyset) = \{\emptyset\}$

     c. $\mathscr{P}(\mathscr{P}(\mathscr{P}(\emptyset))) = \{\emptyset, \{\emptyset\}, \{\{\emptyset\}\}, \{\emptyset, \{\emptyset\}\}\}$

Section 6.2

6. (2) **a.** $x \in A \cap (B \cup C)$ **b.** or   **c.** and   **d.** $A \cap (B \cup C)$ **e.** by definition of intersection, $x \in A$ and $x \in C$. Since $x \in C$, then, by definition of union, $x \in B \cup C$. Hence $x \in A$ and $x \in B \cup C$, and so, by definition of intersection, $x \in A \cap (B \cup C)$.

   (3) **a.** $A \cap (B \cup C) = (A \cap B) \cup (A \cap C)$

9. The proof that $(A - B) \cup (C - B) \subseteq (A \cup C) - B$ is in Appendix B.

   Proof that $(A \cup C) - B \subseteq (A - B) \cup (C - B)$:

   Suppose that x is any element in $(A \cup C) - B$. *[We must show that $x \in (A - B) \cup (C - B)$.]*

   By definition of set difference, $x \in (A \cup C)$ and $x \notin B$.

   And, by definition of union, $x \in A$ or $x \in C$, and in both cases, $x \notin B$.

   Case 1 $(x \in A$ and $x \notin B)$: Then, by definition of set difference, $x \in A - B$, and so by definition of union, $x \in (A - B) \cup (C - B)$.

   Case 2 $(x \in C$ and $x \notin B)$: Then, by definition of set difference, $x \in C - B$, and so by definition of union, $x \in (A - B) \cup (C - B)$.

   In both cases, $x \in (A - B) \cup (C - B)$ *[as was to be shown]*.

   So $(A \cup C) - B \subseteq (A - B) \cup (C - B)$.

   Because both subset containments have now been proved (one here and the other in Appendix B), we conclude that $(A - B) \cup (C - B) = (A \cup C) - B$.

15. Proof:

   Suppose A and B are sets and $A \subseteq B$. Let $x \in B^c$. *[We must show that $x \in A^c$.]*

   By definition of complement, $x \notin B$.

   It follows that $x \notin A$ *[because if $x \in A$ then $x \in B$ (since $A \subseteq B$), and this would contradict the fact that $x \notin B$]*.

   Hence by definition of complement $x \in A^c$.

   [Thus $B^c \subseteq A^c$ by definition of subset.]

21. The "proof" claims that because $x \notin A$ or $x \notin B$, it follows that $x \notin A \cup B$. But it is possible for "$x \notin A$ or $x \notin B$" to be true and "$x \notin A \cup B$" to be false. For example, let $A = \{1, 2\}$, $B = \{2, 3\}$, and $x = 3$. Then since $3 \notin \{1, 2\}$, the statement "$x \notin A$ or $x \notin B$" is true. But since $A \cup B = \{1, 2, 3\}$ and $3 \in \{1, 2, 3\}$, the statement "$x \notin A \cup B$" is false.

36. *a.* Proof:

   Let A and B be any sets. *[We will show that $[(A - B) \cup (B - A) \cup (A \cap B] \subseteq A \cup B$ and that $A \cup B \subseteq (A - B) \cup (B - A) \cup (A \cap B)$.]*

   Proof that $(A - B) \cup (B - A) \cup (A \cap B \subseteq A \cup B$:

   Suppose $x \in (A - B) \cup (B - A) \cup (A \cap B)$. *[We must show that $x \in A \cup B$.]*

   By definition of union, $x \in A - B$ or $x \in B - A$ or $x \in A \cap B$.

   Case 1 $(x \in A - B)$: In this case, by definition of set difference, $x \in A$ and $x \notin B$. In particular, $x \in A$. Then, by definition of union, $x \in A \cup B$.

   Case 2 $(x \in B - A)$: In this case, by definition of set difference, $x \in B$ and $x \notin A$. In particular, $x \in B$. Then, by definition of union, $x \in A \cup B$.

   Case 3 $(x \in A \cap B)$: In this case, by definition of intersection, $x \in A$ and $x \in B$. Then, by definition of union, $x \in A \cup B$.

   In all three cases, $x \in A \cup B$ *[as was to be shown]*.

Proof that $A \cup B \subseteq (A - B) \cup (B - A) \cup (A \cap B)$:

Suppose $x \in A \cup B$. *[We must show that $x \in (A - B) \cup (B - A) \cup (A \cap B)$.]*

By definition of union, $x \in A$ or $x \in B$.

Case 1 ($x \in A$): In this case, either $x \in B$ or $x \notin B$. If $x \in B$, then, since x is also in A, $x \in A \cap B$ by definition of intersection. It follows by definition of union that $x \in (A - B) \cup (B - A) \cup (A \cap B)$. If $x \notin B$, then, since x is also in A, $x \in A - B$ by definition of set difference. It follows by definition of union that $x \in (A - B) \cup (B - A) \cup (A \cap B)$.

Case 2 ($x \in B$): In this case, either $x \in A$ or $x \notin A$. If $x \in A$, then, since x is also in B, $x \in A \cap B$ by definition of intersection. It follows by definition of union that $x \in (A - B) \cup (B - A) \cup (A \cap B)$. If $x \notin A$, then, since x is also in B, $x \in B - A$ by definition of set difference. It follows by definition of union that $x \in (A - B) \cup (B - A) \cup (A \cap B)$.

In both cases, $x \in (A - B) \cup (B - A) \cup (A \cap B)$ *[as was to be shown]*.

[Since both subset containments have been proved, $(A - B) \cup (B - A) \cup (A \cap B) = A \cup B$ by definition of set equality.]

39. <u>Proof:</u>

Let $A_1, A_2, \ldots, A_n$ and B be any sets. *[We will show that $\bigcap_{i=1}^{n}(A_i - B) = \left(\bigcap_{i=1}^{n}A_i\right) - B$.]*

Proof that $\bigcap_{i=1}^{n}(A_i - B) \subseteq \left(\bigcap_{i=1}^{n}A_i\right) - B$:

Suppose $x \in \bigcap_{i=1}^{n}(A_i - B)$. *[Show that $x \in \left(\bigcap_{i=1}^{n}A_i\right) - B$.]*

By definition of general intersection, $x \in A_i - B$ for all integers $i = 1, 2, \ldots, n$.

And by definition of set difference, $x \in A_i$ and $x \notin B$ for all integers $i = 1, 2, \ldots, n$.

It follows by definition of general intersection that $x \in \left(\bigcap_{i=1}^{n}A_i\right)$, and it is also the case that $x \notin B$.

Hence $x \in \left(\bigcap_{i=1}^{n}A_i\right) - B$ by definition of set difference *[as was to be shown]*.

Proof that $\left(\bigcap_{i=1}^{n}A_i\right) - B \subseteq \bigcap_{i=1}^{n}(A_i - B)$:

Suppose $x \in \left(\bigcap_{i=1}^{n}A_i\right) - B$. *[Show that $x \in \bigcap_{i=1}^{n}(A_i - B)$.]*

By definition of set difference, $x \in \left(\bigcap_{i=1}^{n}A_i\right)$ and $x \notin B$.

And by definition of general intersection, $x \in A_i$ for all integers $i = 1, 2, \ldots, n$, and it is also the case that $x \notin B$.

Hence $x \in (A_i - B)$ for all integers $i = 1, 2, \ldots, n$.

So $x \in \bigcap_{i=1}^{n}(A_i - B)$ by definition of general intersection *[as was to be shown]*.

[Since both subset containments have been proved, $\bigcap_{i=1}^{n}(A_i - B) = \left(\bigcap_{i=1}^{n}A_i\right) - B$ by definition of set equality.]

Section 6.3

12. True. <u>Proof</u>: Let A, B, and C be any sets. *[We must show that $A \cap (B-C) = (A \cap B) - (A \cap C)$.]*

    Proof that $A \cap (B - C) \subseteq (A \cap B) - (A \cap C)$:

    Suppose $x \in A \cap (B - C)$. *[We must show that $x \in (A \cap B) - (A \cap C)$.]*

    By definition of intersection, $x \in A$ and $x \in (B - C)$, and so

    $x \in A$ and, by definition of set difference, $x \in B$ and $x \notin C$.

    Now if x were in $A \cap C$, then x would be in C, which it is not.

    Thus $x \notin A \cap C$, and so $x \in A \cap B$ and $x \notin A \cap C$.

    Hence $x \in (A \cap B) - (A \cap C)$ by definition of set difference *[as was to be shown]*.

    [Therefore, $A \cap (B - C) \subseteq (A \cap B) - (A \cap C)$.]

    Proof that $(A \cap B) - (A \cap C) \subseteq A \cap (B - C)$:

    Suppose $x \in (A \cap B) - (A \cap C)$. *[We must show that $x \in A \cap (B - C)$.]*

    By definition of set difference, $x \in A \cap B$ and $x \notin A \cap C$, and so,

    by definition of intersection, $x \in A$ and $x \in B$, and also $x \notin A \cap C$.

    Now if x were in C then x would be in both A and C, and so x would be in $A \cap C$ which it is not.

    Thus $x \in A$ and $x \in B$ and $x \notin C$, and hence

    $x \in A$ and $x \in B - C$ by definition of set difference.

    Finally, by definition of intersection, $x \in A \cap (B - C)$ *[as was to be shown]*.

    [Therefore, $(A \cap B) - (A \cap C) \subseteq A \cap (B - C)$.]

    [Since both subset containments have been proved, $A \cap (B-C) = (A \cap B) - (A \cap C)$ by definition of set equality.]

15. True. <u>Proof</u>: Let A, B, and C be any sets such that $A \cap C = B \cap C$ and $A \cup C = B \cup C$. *[We must show that $A = B$.]*

    Proof that $A \subseteq B$

    Suppose $x \in A$. *[We must show that $x \in B$.]*

    Either $x \in C$ or $x \notin C$.

    Case 1 ($x \in C$): In this case, $x \in A$ and $x \in C$, and so, by definition of intersection, $x \in A \cap C$. But $A \cap C = B \cap C$ by hypothesis, and hence $x \in B \cap C$ by definition of subset. Thus, in particular, $x \in B$ *[as was to be shown]*.

    Case 2 ($x \notin C$): Since $x \in A$, by definition of union, $x \in A \cup C$. Now, by hypothesis, $A \cup C = B \cup C$. So $x \in B \cup C$, and, by definition of union, $x \in B$ or $x \in C$. But in this case $x \notin C$, and so $x \in B$ *[as was to be shown]*.

    [Therefore, $A \subseteq B$ by definition of subset.]

    Proof that $B \subseteq A$

    Suppose $x \in B$. *[We must show that $x \in A$.]*

    Either $x \in C$ or $x \notin C$.

    Case 1 ($x \in C$): In this case, $x \in B$ and $x \in C$, and so, by definition of intersection, $x \in B \cap C$. But $B \cap C = A \cap C$ by hypothesis, and hence $x \in A \cap C$ by definition of subset. Thus, in particular, $x \in A$ *[as was to be shown]*.

    Case 2 ($x \notin C$): Since $x \in B$, by definition of union, $x \in B \cup C$. Now, by hypothesis, $B \cup C = A \cup C$. So $x \in A \cup C$, and, by definition of union, $x \in A$ or $x \in C$. But in this case $x \notin C$, and so $x \in A$ *[as was to be shown]*.

[Therefore, $B \subseteq A$ by definition of subset.]

[Since both subset containments have been proved, $A = B$ by definition of set equality.]

21. False. The elements of $\mathscr{P}(A \times B)$ are subsets of $A \times B$, whereas the elements of $\mathscr{P}(A) \times \mathscr{P}(B)$ are ordered pairs whose first element is a subset of A and whose second element is a subset of B.

    Counterexample: Let $A = B = \{1\}$.

    Then $\mathscr{P}(A) = \{\emptyset, \{1\}\}$, $\mathscr{P}(B) = \{\emptyset, \{1\}\}$, and $\mathscr{P}(A) \times \mathscr{P}(B) = \{(\emptyset, \emptyset), (\emptyset, \{1\}), (\{1\}, \emptyset), (\{1\}, \{1\})\}$.

    On the other hand, $A \times B = \{(1, 1)\}$, and so $\mathscr{P}(A \times B) = \{\emptyset, \{(1, 1)\}\}$.

    By inspection $\mathscr{P}(A) \times \mathscr{P}(B) \neq \mathscr{P}(A \times B)$.

24. No. The sets S_a, S_b, S_c, and $S_\emptyset$ do not form a partition of $\mathscr{P}(S)$ because they are not mutually disjoint. For example, $\{a, b\} \in S_a$ and $\{a, b\} \in S_b$.

33. Proof: Let A and B be any sets. Then

$$
\begin{aligned}
(A - B) \cap (A \cap B) &= (A \cap B^c) \cap (A \cap B) && \text{by the set difference law} \\
&= A \cap [B^c \cap (A \cap B)] && \text{by the associative law for } \cap \\
&= A \cap [(A \cap B) \cap B^c] && \text{by the commutative law for } \cap \\
&= A \cap [A \cap (B \cap B^c)] && \text{by the associative law for } \cap \\
&= A \cap [A \cap \emptyset] && \text{by the complement law for } \cap \\
&= A \cap \emptyset && \text{by the identity law for } \cap \\
&= \emptyset && \text{by the identity law for } \cap.
\end{aligned}
$$

39. Proof: Let A and B be any sets. Then

$(A - B) \cup (B - A)$
$$
\begin{aligned}
&= (A \cap B^c) \cup (B \cap A^c) && \text{by the set difference law (used twice)} \\
&= [(A \cap B^c) \cup B] \cap [(A \cap B^c) \cup A^c] && \text{by the distributive law} \\
&= [B \cup (A \cap B^c)] \cap [A^c \cup (A \cap B^c)] && \text{by the commutative law for } \cup \text{ (used twice)} \\
&= [(B \cup A) \cap (B \cup B^c)] \cap [(A^c \cup A) \cap (A^c \cup B^c)] && \text{by the distributive law (used twice)} \\
&= [(A \cup B) \cap (B \cup B^c)] \cap [(A \cup A^c) \cap (A^c \cup B^c)] && \text{by the commutative law for } \cup \text{ (used twice)} \\
&= [(A \cup B) \cap U] \cap [U \cap (A^c \cup B^c)] && \text{by the complement law for } \cup \text{ (used twice)} \\
&= [(A \cup B) \cap U] \cap [(A^c \cup B^c) \cap U] && \text{by the commutative law for } \cap \\
&= (A \cup B) \cap (A^c \cup B^c) && \text{by the identity law for } \cap \text{ (used twice)} \\
&= (A \cup B) \cap (A \cap B)^c && \text{by De Morgan's law} \\
&= (A \cup B) - (A \cap B) && \text{by the set difference law.}
\end{aligned}
$$

42. Let A and B be any sets. Then

$(A - (A \cap B)) \cap (B - (A \cap B))$
$$
\begin{aligned}
&= (A \cap (A \cap B)^c) \cap (B \cap (A \cap B)^c) && \text{by the set difference law (used twice)} \\
&= A \cap ((A \cap B)^c \cap (B \cap (A \cap B)^c)) && \text{by the associative law for } \cap \\
&= A \cap (((A \cap B)^c \cap B) \cap (A \cap B)^c) && \text{by the associative law for } \cap \\
&= A \cap ((B \cap (A \cap B)^c) \cap (A \cap B)^c) && \text{by the commutative law for } \cap \\
&= A \cap (B \cap ((A \cap B)^c \cap (A \cap B)^c)) && \text{by the associative law for } \cap \\
&= A \cap (B \cap (A \cap B)^c) && \text{by the idempotent law for } \cap \\
&= (A \cap B) \cap (A \cap B)^c && \text{by the associative law for } \cap \\
&= \emptyset && \text{by the complement law for } \cap.
\end{aligned}
$$

45. *a.* Proof: Let A, B, and C be any sets.

    Proof that $(A - B) \cup (B - C) \subseteq (A \cup B) - (B \cap C)$: Suppose $x \in (A - B) \cup (B - C)$. By definition of union, $x \in A - B$ or $x \in B - C$.

    Case 1 ($x \in A - B$): In this case, by definition of set difference, $x \in A$ and $x \notin B$. Then since $x \in A$, by definition of union, $x \in A \cup B$. Also, since $x \notin B$, then $x \notin B \cap C$ (for otherwise, by definition of intersection, x would be in B, which it is not). Thus $x \in A \cup B$ and $x \notin B \cap C$, and so, by definition of set difference, $x \in (A \cup B) - (B \cap C)$.

Case 2 ($x \in B - C$): In this case, by definition of set difference, $x \in B$ and $x \notin C$. Then since $x \in B$, by definition of union, $x \in A \cup B$. Also, since $x \notin C$, then $x \notin B \cap C$ (for otherwise, by definition of intersection, x would be in C, which it is not). Thus $x \in A \cup B$ and $x \notin B \cap C$, and so, by definition of set difference, $x \in (A \cup B) - (B \cap C)$.

Hence, in both cases, $x \in (A \cup B) - (B \cap C)$, and so, by definition of subset,

$$(A - B) \cup (B - C) \subseteq (A \cup B) - (B \cap C).$$

Proof that $(A \cup B) - (B \cap C) \subseteq (A - B) \cup (B - C)$: Suppose $x \in (A \cup B) - (B \cap C)$. By definition of set difference, $x \in A \cup B$ and $x \notin B \cap C$. Note that either $x \in B$ or $x \notin B$.

Case 1 ($x \in B$): In this case $x \notin C$ because otherwise x would be in both B and C, which would contradict the fact that $x \notin B \cap C$. Thus, in this case, $x \in B$ and $x \notin C$, and so $x \in B - C$ by definition of set difference. Then $x \in (A - B) \cup (B - C)$ by definition of union.

Case 2 ($x \notin B$): In this case, since $x \in A \cup B$, then $x \in A$. Hence $x \in A$ and $x \notin B$, and so $x \in A - B$ by definition of set difference. Then $x \in (A - B) \cup (B - C)$ by definition of union.

Hence, in both cases, $x \in (A - B) \cup (B - C)$, and so, by definition of subset,

$$(A \cup B) - (B \cap C) \subseteq (A - B) \cup (B - C).$$

Therefore, since both set containments have been proved, we conclude that

$$(A - B) \cup (B - C) = (A \cup B) - (B \cap C)$$

by definition of set equality.

b. <u>Proof</u>: Let A, B, and C be any sets. Then

$(A - B) \cup (B - C)$

$=$	$(A \cap B^c) \cup (B \cap C^c)$	by the set difference law (used twice)
$=$	$((A \cap B^c) \cup B) \cap ((A \cap B^c) \cup C^c)$	by the distributive law
$=$	$(B \cup (A \cap B^c)) \cap ((A \cap B^c) \cup C^c)$	by the commutative law for $\cup$
$=$	$((B \cup A) \cap (B \cup B^c)) \cap ((A \cap B^c) \cup C^c)$	by the distributive law
$=$	$((B \cup A) \cap U) \cap ((A \cap B^c) \cup C^c)$	by the complement law for $\cup$
$=$	$(B \cup A) \cap ((A \cap B^c) \cup C^c)$	by the identity law for $\cap$
$=$	$(A \cup B) \cap ((A \cap B^c) \cup C^c)$	by the commutative law for $\cup$
$=$	$((A \cup B) \cap (A \cap B^c)) \cup ((A \cup B) \cap C^c)$	by the distributive law
$=$	$(((A \cup B) \cap A) \cap B^c) \cup ((A \cup B) \cap C^c)$	by the associative law for $\cap$
$=$	$((A \cap (A \cup B)) \cap B^c) \cup ((A \cup B) \cap C^c)$	by the commutative law for $\cap$
$=$	$(A \cap B^c) \cup ((A \cup B) \cap C^c)$	by the absorption law
$=$	$((A \cap B^c) \cup \emptyset) \cup ((A \cup B) \cap C^c)$	by the identity law for $\cup$
$=$	$((A \cap B^c) \cup (B \cap B^c)) \cup ((A \cup B) \cap C^c)$	by the complement law for $\cap$
$=$	$((B^c \cap A) \cup (B^c \cap B)) \cup ((A \cup B) \cap C^c)$	by the commutative law for $\cap$
$=$	$(B^c \cap (A \cup B)) \cup ((A \cup B) \cap C^c)$	by the distributive law
$=$	$((A \cup B) \cap B^c)) \cup ((A \cup B) \cap C^c)$	by the commutative law for $\cap$
$=$	$(A \cup B) \cap (B^c \cup C^c)$	by the distributive law
$=$	$(A \cup B) \cap (B \cap C)^c$	by De Morgan's law
$=$	$(A \cup B) - (B \cap C)$	by the set difference law.

c. Although writing down every detail of the element proof is somewhat tedious, its basic idea is not hard to see. In this case the element proof is probably easier than the algebraic proof.

51. <u>Lemma</u>: For any subsets A and B of a universal set U and for any element x,

    (1) $x \in A \triangle B \Leftrightarrow (x \in A \text{ and } x \notin B) \text{ or } (x \notin A \text{ and } x \in B)$

    (2) $x \notin A \triangle B \Leftrightarrow (x \notin A \text{ and } x \notin B) \text{ or } (x \in A \text{ and } x \in B)$.

<u>Proof</u>:

(1) Suppose A and B are any sets and x is any element. Then

$$
\begin{array}{lll}
x \in A \triangle B & \Leftrightarrow \quad x \in (A - B) \cup (B - A) & \text{by definition of } \triangle \\
& \Leftrightarrow \quad x \in A - B \text{ or } x \in B - A & \text{by definition of } \cup \\
& \Leftrightarrow \quad (x \in A \text{ and } x \notin B) \text{ or } (x \in B \text{ and } x \notin A) & \text{by definition of set difference.}
\end{array}
$$

(2) Suppose A and B are any sets and x is any element.

Observe that there are only four mutually exclusive possibilities for the relationship of x to A and B: $(x \in A \text{ and } x \notin B) \text{ or } (x \in B \text{ and } x \notin A) \text{ or } (x \in A \text{ and } x \in B) \text{ or } (x \notin A \text{ and } x \notin B)$.

By part (1), the condition that $x \in A \triangle B$ is equivalent to the first two possibilities. So the condition that $x \notin A \triangle B$ is equivalent to the second two possibilities.

In other words, $x \notin A \triangle B \Leftrightarrow (x \notin A \text{ and } x \notin B) \text{ or } (x \in A \text{ and } x \in B)$.

<u>Theorem</u>: For all subsets A, B, and C of a universal set U, if $A \triangle C = B \triangle C$ then $A = B$.

<u>Proof</u>: Let A, B, and C be any subsets of a universal set U, and suppose that $A \triangle C = B \triangle C$. *[We will show that $A = B$.]*

Proof that $A \subseteq B$: Suppose $x \in A$. Either $x \in C$ or $x \notin C$. If $x \in C$, then $x \in A$ and $x \in C$ and so by the lemma, $x \notin A \triangle C$. But $A \triangle C = B \triangle C$. Thus $x \notin B \triangle C$ either. Hence, again by the lemma, since $x \in C$ and $x \notin B \triangle C$, then $x \in B$. On the other hand, if $x \notin C$, then by the lemma, since $x \in A$, $x \in A \triangle C$. But $A \triangle C = B \triangle C$. So, again by the lemma, since $x \notin C$ and $x \in B \triangle C$, then $x \in B$. Hence in either case, $x \in B$ *[as was to be shown]*.

Proof that $B \subseteq A$: The proof is exactly the same as for $A \subseteq B$ with the letters A and B interchanged.

Since $A \subseteq B$ and $B \subseteq A$, by definition of set equality $A = B$.

54. <u>Proof</u>:

Suppose A and B are any subsets of a universal set U.

By the universal bound law for $\cap$, $B \cap \emptyset = \emptyset$, and so, by the commutative law for $\cap$, $\emptyset \cap B = \emptyset$.

Take the union with A of both sides to obtain $A \cup (\emptyset \cap B) = A \cup \emptyset$.

But the left-hand side of this equation is $A \cup (\emptyset \cap B) = (A \cup \emptyset) \cap (A \cup B) = A \cap (A \cup B)$ by the distributive law and the identity law for $\cup$.

And the right-hand side of the equation equals A by the by the identity law for $\cup$.

Hence $A \cap (A \cup B) = A$ *[as was to be shown]*.

Section 6.4

3. *a.* commutative law for $\cdot$ *b.* distributive law for $\cdot$ over $+$

    c. idempotent law for $\cdot$ (exercise 1)   *d.* identity law for $\cdot$

    e. distributive law for $\cdot$ over $+$ *f.* commutative law for $+$ *g.* identity law for $\cdot$

Note that once Theorem 6.4.1(5b) has been proved (exercise 4), the proof of this property (Theorem 6.4.1(7a)) can be streamlined as shown below.

<u>Proof</u>: For all elements a and b in B,

$$
\begin{array}{lll}
(a + b) \cdot a & = & (a + b) \cdot (a + 0) & \text{by the identity law for } + \\
& = & a + (b \cdot 0) & \text{by the distributive law for } + \text{ over } \cdot \\
& = & a + 0 & \text{by exercise 4} \\
& = & a & \text{by the identity law for } +.
\end{array}
$$

6. *b.* <u>Proof</u>: By the uniqueness of the complement law, to show that $\overline{1} = 0$, it suffices to show that $1 + 0 = 1$ and $1 \cdot 0 = 0$. But the first equation is true by the identity law for $+$, and the second equation is true by exercise 4 (the universal bound law for $\cdot$). Thus $\overline{1} = 0$.

9. <u>Proof 1</u>: By exercise 8, we know that for all x and y in B, $\overline{x \cdot y} = \overline{x} + \overline{y}$. So suppose a and b are any elements in B. Substitute $\overline{a}$ and $\overline{b}$ in place of x and y in this equation to obtain $\overline{\overline{a} \cdot \overline{b}} = \overline{\overline{a}} + \overline{\overline{b}}$, and since $\overline{\overline{a}} + \overline{\overline{b}} = a + b$ by the double complement law, we have $\overline{\overline{a} \cdot \overline{b}} = a + b$. Hence by the uniqueness of the complement law, the complement of $\overline{a} \cdot \overline{b}$ is $a + b$. It follows by definition of complement that

$$(\overline{a} \cdot \overline{b}) + (a + b) = 1 \qquad \text{and} \qquad (\overline{a} \cdot \overline{b}) \cdot (a + b) = 0.$$

By the commutative laws for $+$ and $\cdot$,

$$(a + b) + (\overline{a} \cdot \overline{b}) = 1 \qquad \text{and} \qquad (a + b) \cdot (\overline{a} \cdot \overline{b}) = 0,$$

and thus by the uniqueness of the complement law, the complement of $a + b$ is $\overline{a} \cdot \overline{b}$. In other words, $\overline{a + b} = \overline{a} \cdot \overline{b}$.

<u>Proof 2</u>: An alternative proof can be obtained by taking the proof for exercise 8 in Appendix B and changing every $+$ sign to a $\cdot$ sign and every $\cdot$ sign to a $+$ sign.

12. To avoid unneeded parentheses, assume that $\cdot$ takes precedence over $+$.

<u>Lemma 1</u>: The universal bound laws for a Boolean algebra can be derived from the Boolean algebra axioms without using the associative laws.

<u>Proof</u>: Suppose a is any element of a Boolean algebra B.

$$
\begin{aligned}
a + 1 &= (a+1) \cdot 1 & &\text{because 1 is an identity for } \cdot \\
&= (a+1) \cdot (a + \overline{a}) & &\text{by the complement law for } + \\
&= a + 1 \cdot \overline{a} & &\text{by the distributive law for } + \text{ over } \cdot \\
&= a + \overline{a} \cdot 1 & &\text{by the commutative law for } \cdot \\
&= a + \overline{a} & &\text{because 1 is an identity for } \cdot \\
&= 1 & &\text{by the complement law for } +.
\end{aligned}
$$

The proof that $a \cdot 0 = 0$ is obtained using the same sequence of steps but changing each $\cdot$ to $+$ and each $+$ to $\cdot$.

<u>Lemma 2</u>: The absorption laws for a Boolean algebra can be derived from the Boolean algebra axioms without using the associative laws.

<u>Proof</u>: Suppose a and b are any elements of a Boolean algebra B.

$$
\begin{aligned}
a \cdot b + a &= a \cdot b + a \cdot 1 & &\text{because 1 is an identity for } \cdot \\
&= a \cdot (b + 1) & &\text{by the distributive law for } \cdot \text{ over } + \\
&= a \cdot 1 & &\text{by the universal bound law for } + \text{ (Lemma 1)} \\
&= a & &\text{because 1 is an identity for } \cdot.
\end{aligned}
$$

The proof that $a \cdot (a + b) = a$ is obtained using the same sequence of steps but changing each $\cdot$ to $+$ and each $+$ to $\cdot$.

Note also that the proofs of the idempotent laws (Example 6.4.2 and the solution to exercise 1) use only the Boolean algebra axioms without the associative laws.

<u>Theorem 1</u>: The associative law for $+$ can be derived from the other axioms in a Boolean algebra.

<u>Proof</u>:

Part 1: We first prove that for all x, y, and z in B, (1) $(x + (y + z)) \cdot x = x$ and (2) $((x + y) + z) \cdot x = x$.

So suppose x, y, and z are any elements in B. It follows immediately from an absorption law that

$$(1)\ (x + (y + z)) \cdot x = x$$

Also,

$$
\begin{aligned}
((x + y) + z)) \cdot x
&= x \cdot ((x + y) + z) && \text{by the commutative law for } \cdot \\
&= x \cdot (x + y) + x \cdot z && \text{by the distributive law for } \cdot \text{ over } + \\
&= (x + y) \cdot x + x \cdot z && \text{by the commutative law for } \cdot \\
&= x + x \cdot z && \text{by an absorption law} \\
&= x \cdot x + x \cdot z && \text{by the idempotent law for } \cdot \\
&= x \cdot (x + z) && \text{by the distributive law for } \cdot \text{ over } + \\
&= (x + z) \cdot x && \text{by the commutative law for } \cdot \\
&= x && \text{by an absorption law.}
\end{aligned}
$$

Hence

$$(2)\ ((x + y) + z) \cdot x = x.$$

Part 2: By the commutative law for $+$ and equation (2), for all x, y, and z in B,

$$((x + y) + z) \cdot y = ((y + x) + z) \cdot y = y.$$

And by the commutative law for $+$ and equation (2), for all x, y, and z in B,

$$(x + (y + z)) \cdot y = ((y + x) + z) \cdot y = y.$$

Thus we have

$$(3)\ ((x + y) + z) \cdot y = y \quad \text{and} \quad (4)\ (x + (y + z)) \cdot y = y.$$

By similar reasoning we can also conclude that

$$(5)\ ((x + y) + z) \cdot z = z \quad \text{and} \quad (6)\ (x + (y + z)) \cdot z = z.$$

Part 3: We next prove that for all a, b, and c in B,

$$(7)\ a + (b + c) = ((a + b) + c) \cdot (a + (b + c)) \quad \text{and} \quad (8)\ (a + b) + c = ((a + b) + c) \cdot (a + (b + c)).$$

To prove (7), suppose a, b, and c are any elements in B. Then
$((a + b) + c) \cdot (a + (b + c))$

$$
\begin{aligned}
&= ((a + b) + c) \cdot a + ((a + b) + c) \cdot (b + c) && \text{by the distributive law for } \cdot \text{ over } + \\
&= a + ((a + b) + c) \cdot (b + c) && \text{by equation (2)} \\
&= a + [((a + b) + c) \cdot b + ((a + b) + c) \cdot c] && \text{by the distributive law for } \cdot \text{ over } + \\
&= a + (b + c) && \text{by equations (3) and (5).}
\end{aligned}
$$

Similarly, if a, b, and c are any elements in B. Then we can prove equation (8) as follows:
$((a + b) + c) \cdot (a + (b + c))$

$$
\begin{aligned}
&= (a + (b + c)) \cdot ((a + b) + c) && \text{by the commutative law for } \cdot \\
&= (a + (b + c)) \cdot (a + b) + (a + (b + c)) \cdot c && \text{by the distributive law for } \cdot \text{ over } + \\
&= (a + (b + c)) \cdot (a + b) + c && \text{by equation (6)} \\
&= [(a + (b + c)) \cdot a + (a + (b + c)) \cdot b] + c && \text{by the distributive law for } \cdot \text{ over } + \\
&= (a + b) + c && \text{by equations (1) and (4).}
\end{aligned}
$$

Therefore, since both $a + (b + c)$ and $(a + b) + c$ are equal to the same quantity, they are equal to each other: $a + (b + c) = (a + b) + c$.

<u>Theorem 2</u>: The associative law for $\cdot$ can be derived from the other axioms in a Boolean algebra.

<u>Proof</u>: Suppose a, b, and c are any elements in a Boolean algebra B. The proof that $(a \cdot b) \cdot c = a \cdot (b \cdot c)$ is obtained using the same sequence of steps as in the proof of Theorem 1 but changing each $\cdot$ to $+$ and each $+$ to $\cdot$.

Alternative (Shorter) Proof for Theorem 1 (based on the outline in *Introduction to Boolean Algebra* by S. Givant and P. Halmos, Springer, 2010)

The Cancellation Law for $\cdot$: For all elements p, q, and r in a Boolean algebra,

$$\text{if } q \cdot p = r \cdot p \text{ and } q \cdot \overline{p} = r \cdot \overline{p}, \text{ then } q = r.$$

<u>Lemma 3</u>: The cancellation law for $\cdot$ can be derived from the other axioms in a Boolean algebra without using the associative law.

<u>Proof</u>: Suppose p, q, and r are any elements in a Boolean algebra B such that

$$q \cdot p = r \cdot p \quad \text{and} \quad q \cdot \overline{p} = r \cdot \overline{p}. \qquad (1)$$

We will show that $q = r$.

Now

$$
\begin{aligned}
q \cdot p + q \cdot \overline{p} &= q \cdot (p + \overline{p}) && \text{by the distributive law for } \cdot \text{ over } + \\
&= q \cdot 1 && \text{by the complement law for } + \\
&= q && \text{by the identity law for 1.}
\end{aligned}
$$

Similarly,

$$
\begin{aligned}
r \cdot p + r \cdot \overline{p} &= r \cdot (p + \overline{p}) && \text{by the distributive law for } \cdot \text{ over } + \\
&= r \cdot 1 && \text{by the complement law for } + \\
&= r && \text{by the identity law for 1.}
\end{aligned}
$$

But, by substitution from (1),

$$q \cdot p + q \cdot \overline{p} = r \cdot p + r \cdot \overline{p}.$$

Thus

$$q = r.$$

<u>Theorem 1</u>: The associative law for $+$ can be derived from the other axioms in a Boolean algebra.

<u>Proof</u>: Suppose a, b, and c are any elements in a Boolean algebra B.

Part 1: We first prove that $(a + (b + c)) \cdot a = ((a + b) + c) \cdot a$.

$$(a + (b + c)) \cdot a = a \quad \text{by an absorption law (Lemma 2).}$$

In addition:

$$
\begin{aligned}
((a + b) + c) \cdot a &= a \cdot ((a + b) + c) && \text{by the commutative law for } \cdot \\
&= a \cdot (a + b) + a \cdot c && \text{by the distributive law for } \cdot \text{ over } + \\
&= (a + b) \cdot a + a \cdot c && \text{by the commutative law for } + \\
&= a + a \cdot c && \text{by an absorption law} \\
&= a \cdot c + a && \text{by the commutative law for } + \\
&= a && \text{by an absorption law.}
\end{aligned}
$$

Since both quantities equal a, we conclude that

$$(a + (b + c)) \cdot a = ((a + b) + c) \cdot a.$$

Part 2: We next prove that $(a + (b + c)) \cdot \overline{a} = ((a + b) + c) \cdot \overline{a}$

$$
\begin{aligned}
(a + (b + c)) \cdot \overline{a} &= \overline{a} \cdot (a + (b + c)) && \text{by the commutative law for } \cdot \\
&= \overline{a} \cdot a + \overline{a} \cdot (b + c) && \text{by the distributive law for } \cdot \text{ over } + \\
&= 0 + \overline{a} \cdot (b + c) && \text{by the complement law for } \cdot \\
&= \overline{a} \cdot (b + c) + 0 && \text{by the commutative law for } + \\
&= \overline{a} \cdot (b + c) && \text{because 0 is an identity for } +.
\end{aligned}
$$

In addition:

$$
\begin{aligned}
((a + b) + c) \cdot \overline{a} &= \overline{a} \cdot ((a + b) + c) && \text{by the commutative law for } \cdot \\
&= \overline{a} \cdot (a + b) + \overline{a} \cdot c && \text{by the distributive law for } \cdot \text{ over } + \\
&= (\overline{a} \cdot a + \overline{a} \cdot b) + \overline{a} \cdot c && \text{by the distributive law for } \cdot \text{ over } + \\
&= (a \cdot \overline{a} + \overline{a} \cdot b) + \overline{a} \cdot c && \text{by the commutative law for } \cdot \\
&= (0 + \overline{a} \cdot b) + \overline{a} \cdot c && \text{by the complement law for } \cdot \\
&= (\overline{a} \cdot b + 0) + \overline{a} \cdot c && \text{by the commutative law for } + \\
&= \overline{a} \cdot b + \overline{a} \cdot c && \text{because 0 is an identity for } + \\
&= \overline{a} \cdot (b + c) && \text{by the distributive law for } \cdot \text{ over } +.
\end{aligned}
$$

Since both quantities equal $\overline{a} \cdot (b + c)$, we conclude that

$$
(a + (b + c)) \cdot \overline{a} = ((a + b) + c) \cdot \overline{a}.
$$

Part 3: Parts (1) and (2) show that

$$
(a + (b + c)) \cdot a = ((a + b) + c) \cdot a \quad \text{and} \quad (a + (b + c)) \cdot \overline{a} = ((a + b) + c) \cdot \overline{a}.
$$

Thus, by the cancellation law

$$
a + (b + c) = (a + b) + c.
$$

15. This statement contradicts itself. If it were true, then because it declares itself to be a lie, it would be false. Consequently, it is not true. On the other hand, if it were false, then it would be false that "the sentence in this box is a lie," and so the sentence would be true. Consequently, the sentence is not false. Thus the sentence is neither true nor false, which contradicts the definition of a statement. Hence the sentence is not a statement.

18. In order for an *and* statement to be true, both components must be true. So if the given sentence is a true statement, the first component "this sentence is false" is true. But this implies that the sentence is false. In other words, the sentence is not true. On the other hand, if the sentence is false, then at least one component is false. But because the second component "$1 + 1 = 2$" is true, the first component must be false. Thus it is false that "this sentence is false," and so the sentence is true. In other words, the sentence is not false. Thus the sentence is neither true nor false, which contradicts the definition of a statement. Hence the sentence is not a statement.

24. Because the total number of strings consisting of 11 or fewer English words is finite, the number of such strings that describe integers must be also finite. Thus the number of integers described by such strings must be finite, and hence there is a largest such integer, say m. Let $n = m + 1$. Then n is "the smallest integer not describable in fewer than 12 English words." But this description of n contains only 11 words. So n is describable in fewer than 12 English words, which is a contradiction. (*Comment*: This contradiction results from the self-reference in the description of n.)

Review Guide: Chapter 6

Definitions and Notation:

- How can you express the definition of subset formally as a universal conditional statement? *(p. 337)*
- What is a proper subset of a set? *(p. 337)*
- How are the definitions of subset and equality of sets related? *(p. 339)*
- What are Venn diagrams? *(p. 340)*
- What are the union, intersection, and difference of sets? *(p. 341)*
- What is the complement of a set? *(p. 341)*
- What is the relation between sets and interval notation? *(p. 342)*
- How are unions and intersections defined for indexed collections of sets? *(p. 343)*
- What does it mean for two sets to be disjoint? *(p. 344)*
- What does it mean for a collection of sets to be mutually disjoint? *(p. 345)*
- What is a partition of a set? *(p. 345)*
- What is the power set of a set? *(p. 346)*
- What is an ordered n-tuple? *(p. 346)*
- What is the Cartesian product of n sets, where $n \geq 2$? *(p. 347)*

Set Theory

- How do you use an element argument to prove that one set is a subset of another set? *(p. 337-338)*
- What is the basic (two-step) method for showing that two sets are equal? *(pp. 339 and 356)*
- How are the procedural versions of set operations used to prove properties of sets? *(p. 352-353)*
- Are you familiar with the set properties in Theorems 6.2.1 and 6.2.2? *(pp. 352 and 354)*
- Why is the empty set a subset of every set? *(p. 362)*
- How is the element method used to show that a set equals the empty set? *(p. 362)*
- How do you find a counterexample for a proposed set identity? *(p. 367)*
- How do you find the number of subsets of a set with a finite number of elements? *(p. 369)*
- What is an "algebraic method" for proving that one set equals another set? *(p. 370-371)*
- What is a Boolean algebra? *(p. 375)*
- How do you deduce additional properties of a Boolean algebra from the properties that define it? *(p. 377)*
- What is Russell's paradox? *(p. 378)*
- What is the Halting Problem? *(p. 379)*

Chapter 7: Functions

The aim of Section 7.1 is to promote a broad view of the function concept and to give you experience with the wide variety of functions that arise in discrete mathematics. Representation of functions by arrow diagrams is emphasized to prepare the way for the discussion of one-to-one and onto functions in Section 7.2.

Section 7.2 focuses on function properties. As you are learning about one-to-one and onto functions in this section, you may need to review the logical principles such as the negation of $\forall$, $\exists$, and if-then statements and the equivalence of a conditional statement and its contrapositive. These logical principles are needed to understand the equivalence of the two forms of the definition of one-to-one and what it means for a function not to be one-to-one or onto.

Sections 7.3 and 7.4 go together in the sense that the relations between one-to-one and onto functions and composition of functions developed in Section 7.3 are used to prove the fundamental theorem about cardinality in Section 7.4. The proofs that a composition of one-to-one functions is one-to-one or that a composition of onto functions is onto (and the related exercises) will test the degree to which you have learned to instantiate mathematical definitions in abstract contexts, apply the method of generalizing from the generic particular in a sophisticated setting, develop mental models of mathematical concepts that are both vivid and generic enough to reason with, and create moderately complex chains of deductions.

When you read Section 7.4, try to see the connections that link Russell's paradox, the halting problem, and the Cantor diagonalization argument.

Section 7.1

3. *b.* False. The definition of function does not allow an element of the domain to be associated to two different elements of the co-domain, but it does allow an element of the co-domain to be the image of more than one element in the domain. For example, let $X = \{1, 2\}$ and $Y = \{a\}$ and define $f\colon X \to Y$ by specifying that $f(1) = f(2) = a$. Then f defines a function from X to Y for which a has two unequal preimages.

6. *b.* Define $F\colon \mathbf{Z}^{nonneg} \to \mathbf{R}$ as follows: for each nonnegative integer n, $F(n) = (-1)^n(2n)$.

9. *d.* $S(5) = 1 + 5 = 6$

    e. $S(18) = 1 + 2 + 3 + 6 + 9 + 18 = 39$

    f. $S(21) = 1 + 3 + 7 + 21 = 32$

12. *c.* $G(3,2) = ((2 \cdot 3 + 1) \bmod 5, (3 \cdot 2 - 2) \bmod 5) = (7 \bmod 5, 4 \bmod 5) = (2, 4)$

    d. $G(1,5) = ((2 \cdot 1 + 1) \bmod 5, 3 \cdot 5 - 2) \bmod 5) = (3 \bmod 5, 13 \bmod 5) = (3, 3)$

18. *b.* $\log_2 1024 = 10$ because $2^{10} = 1024$

    d. $\log_2 1 = 0$ because $2^0 = 1$

    e. $\log_{10} \dfrac{1}{10} = -1$ because $10^{-1} = \dfrac{1}{10}$

    f. $\log_3 3 = 1$ because $3^1 = 3$

    g. $\log_2 2^k = k$ because the exponent to which 2 must be raised to obtain 2^k is k
    Alternative answer: $\log_2 2^k = k$ because $2^k = 2^k$

24. Since $\log_b y = 2$, then $b^2 = y$. Now, by properties of exponents, $(b^2)^1 = y$, and so $log_{b^2}(y) = 1$.

27. *b.* $g(aba) = aba$, $g(bbab) = babb$, $g(b) = b$ The range of g is the set of all strings of a's and b's, which equals S.

100

12. *a.* (i) **F is one-to-one**: Suppose n_1 and n_2 are in **Z** and $F(n_1) = F(n_2)$. *[We must show that $n_1 = n_2$.]* By definition of F, $2 - 3n_1 = 2 - 3n_2$. Subtracting 2 from both sides and dividing by -3 gives $n_1 = n_2$.

(ii). **F is not onto**: Let $m = 0$. Then m is in **Z** but $m \neq F(n)$ for any integer n. *[For if $m = F(n)$ then $0 = 2 - 3n$, and so $3n = 2$ and $n = 2/3$. But $2/3$ is not in **Z**.]*

b. **G is onto**: Suppose y is any element of **R**. *[We must show that there is an element x in **R** such that $G(x) = y$.]*

[Scratch work: If such an x exists, then, by definition of G, $y = 2 - 3x$ and so $3x = 2 - y$, or, equivalently, $x = (2 - y)/3$. Let's check to see if this works.]

Let $x = (2 - y)/3$. Then

$$(1)\ x \in \mathbf{R} \quad \text{and} \quad (2)\ G(x) = G\left(\frac{2 - y}{3}\right) = 2 - 3\left(\frac{2 - y}{3}\right) = 2 - (2 - y) = 2 - 2 + y = y.$$

[This is what was to be shown.]

18. **f is one-to-one**:

<u>Proof</u>: Let x_1 and x_2 be any real numbers other than -1, and suppose that $f(x_1) = f(x_2)$. *[We must show that $x_1 = x_2$.]* By definition of f,

$$\frac{x_1 + 1}{x_1 - 1} = \frac{x_2 + 1}{x_2 - 1}.$$

Cross-multiplying gives

$$(x_1 + 1)(x_2 - 1) = (x_2 + 1)(x_1 - 1) \quad \text{or, equivalently,} \quad x_1 x_2 - x_1 + x_2 - 1 = x_1 x_2 - x_2 + x_1 - 1.$$

Adding $1 - x_1 x_2$ to both sides gives $-x_1 + x_2 = -x_2 + x_1$, or, equivalently, $2x_1 = 2x_2$. Dividing both sides by 2 gives $x_1 = x_2$ *[as was to be shown]*.

24. *a.* **N is not one-to-one**: Let $s_1 = a$ and $s_2 = ab$. Then $N(s_1) = N(s_2) = 1$ but $s_1 \neq s_2$.

27. *a.* **T is one-to-one**: $T(n)$ is the set of all the positive divisors of n. Observe that for all positive integers n, the largest element of $T(n)$ is n because n divides n and no integer larger than n divides n.

So suppose n_1 and n_2 are positive integers and $T(n_1) = T(n_2)$. *[We must show that $n_1 = n_2$.]*

Now $T(n_1)$ is the set of all the positive divisors of n_1 and $T(n_2)$ is the set of all the positive divisors of n_2.

So since $T(n_1) = T(n_2)$, the largest element of $T(n_1)$, namely n_1, is the same as the largest element of $T(n_2)$, namely n_2.

Hence $n_1 = n_2$ *[as was to be shown]*.

b. **T is not onto**: The set $\{2\}$ is a finite subset of positive integers, but there is no positive integer n such that $T(n) = \{2\}$. The reason is that the number 1 divides every positive integer, and so 1 must be an element of $T(n)$ for all positive integers n. But $1 \notin \{2\}$. (There are many other examples that show T is not onto.)

30. *a.* **J is one-to-one**: Suppose (r_1, s_1) and (r_2, s_2) are in **Q** $\times$ **Q** and $J(r_1, s_1) = J(r_2, s_2)$. *[We must show that $(r_1, s_1) = (r_2, s_2)$.]* By definition of J,

$$r_1 + \sqrt{2}s_1 = r_2 + \sqrt{2}s_2 \quad \text{and thus} \quad r_1 - r_2 = \sqrt{2}(s_2 - s_1).$$

Note that both $r_1 - r_2$ and $s_2 - s_1$ are rational because they are differences of rational numbers (exercise 17 of Section 4.2).

Suppose $s_2 - s_1 \neq 0$. Then $\sqrt{2}(s_2 - s_1)$ is a product of a nonzero rational number and an irrational number ($\sqrt{2}$), and so it is irrational (exercise 11 of Section 4.6). As a consequence, the rational number $(r_1 - r_2)$ equals the irrational number ($\sqrt{2}(s_2 - s_1)$). Because this is impossible, the supposition that $s_2 - s_1 \neq 0$ must be false, and therefore $s_2 - s_1 = 0$.

Thus, by substitution, $r_1 - r_2 = \sqrt{2}(s_2 - s_1) = \sqrt{2} \cdot 0 = 0$. So

$$r_1 - r_2 = 0 \quad \text{and} \quad s_2 - s_1 = 0 \quad \text{or, equivalently,} \quad r_1 = r_2 \quad \text{and} \quad s_2 = s_1.$$

Hence $(r_1, s_1) = (r_2, s_2)$ *[as was to be shown]*.

b. ***J is not onto***: We show that J is not onto by giving an example of a real number that is not equal to $J(r, s)$ for any rational numbers r and s. For example, consider the number $\sqrt{3}$ and suppose there were rational numbers r and s such that

$$\sqrt{3} = r + \sqrt{2}s.$$

We will show that this supposition leads logically to a contradiction.]

Case 1 (s = 0): In this case, $\sqrt{3} = r$ where r is a rational number, which contradicts the fact that $\sqrt{3}$ is irrational (exercise 16, Section 4.7).

Case 2 (s $\neq$ 0): In this case,

$$\sqrt{3} - \sqrt{2}s = r$$
$$\Rightarrow \quad 3 + 2s^2 - 2s\sqrt{6} = r^2 \qquad \text{by squaring both sides}$$
$$\Rightarrow \quad -2s\sqrt{6} = r^2 - 3 - 2s^2 \qquad \text{by subtracting } 3 + 2s^2 \text{ from both sides}$$
$$\Rightarrow \quad \sqrt{6} = \frac{r^2 - 3 - 2s^2}{-2s} \qquad \text{by dividing both sides by } -2s.$$

But both $r^2 - 3 - 2s^2$ and $-2s$ are rational numbers because products and differences of rational numbers are rational (exercises 15 and 17, Section 4.2), and $-2s$ is nonzero because it is a product of -2 and s, which are both nonzero numbers (zero product property). Thus $\sqrt{6}$ is a quotient of a rational number and a nonzero rational number, which is rational (by the result of exercise 16 in Section 4.2). But this contradicts the fact that $\sqrt{6}$ is irrational (by the result of exercise 22, Section 4.7).

Conclusion: Since a contradiction was obtained in both cases, we conclude that the supposition is false. That is, there are no rational numbers r and s such that $\sqrt{3} = r + \sqrt{2}s$. Therefore J is not onto.

39. If $f: \mathbf{R} \to \mathbf{R}$ is onto and c is any nonzero real number, then $c \cdot f$ is also onto.

    Proof: Suppose $f: \mathbf{R} \to \mathbf{R}$ is onto and c is any nonzero real number.

    Let y be any element of $\mathbf{R}$. *[We must show that there exists an element x in $\mathbf{R}$ such that $c \cdot f(x) = y$.]*

    Since $c \neq 0$, y/c is a real number, and since f is onto, there is an $x \in \mathbf{R}$ with $f(x) = y/c$.

    Then $y = c \cdot f(x) = (c \cdot f)(x)$. So $c \cdot f$ is onto *[as was to be shown]*.

48. By the result of exercise 12a, F is not onto. Hence it is not a one-to-one correspondence.

51. Because D is not one-to-one, D is not a one-to-one correspondence.

54. By the result of exercise 17, f is one-to-one. f is also onto for the following reason. Given any real number y other than 3, let $x = \dfrac{1}{3-y}$. Then x is a real number (because $y \neq 3$) and

$$f(x) = f(\frac{1}{3-y}) = \frac{3\left(\frac{1}{3-y}\right) - 1}{\frac{1}{3-y}} = \frac{3\left(\frac{1}{3-y}\right) - 1}{\frac{1}{3-y}} \cdot \frac{(3-y)}{(3-y)} = \frac{3 - (3-y)}{1} = 3 - 3 + y = y.$$

This calculation also shows that $f^{-1}(y) = \dfrac{1}{3-y}$ for all real numbers $y \neq 3$.

57. **Algorithm 7.2.1   Checking Whether a Function is One-to-One**

[For a given function F with domain $X = \{a[1], a[2], \ldots, a[n]\}$, this algorithm discovers whether or not F is one-to-one. Initially, answer is set equal to "one-to-one". Then the values of $F(a[i])$ and $F(a[j])$ are systematically compared for indices i and j with $1 \leq i < j \leq n$. If at any point it is found that $F(a[i]) = F(a[j])$ and $a[i] \neq a[j]$, then F is not one-to-one, and so answer is set equal to "not one-to-one" and execution ceases. If after all possible values of i and j have been examined, the value of answer is still "one-to-one", then F is one-to-one.]

Input: *n [a positive integer], $a[1], a[2], \ldots, a[n]$ [a one-dimensional array representing the set X], F [a function with domain X]*

Algorithm Body:

   answer := *"one-to-one"*

   $i := 1$

   while ($i \leq n - 1$ and *answer* = *"one-to-one"*)

      $j := i + 1$

      while ($j \leq n$ and *answer* = *"one-to-one"*)

         if ($F(a[i]) = F(a[j])$ and $a[i] \neq a[j]$) **then** *answer* := *"not one-to-one"*

         $j := j + 1$

      end while

      $i := i + 1$

   end while

Output: *answer [a string]*

Section 7.3

12. *b.* For all positive real numbers b and x, $\log_b x$ is the exponent to which b must be raised to obtain x. So if b is raised to this exponent, x is obtained. In other words, $b^{\log_b x} = x$.

15. *b.* $z/2 = t/2$ *c.* $f(x_1) = f(x_2)$

18. f must be one-to-one.

   Proof:

   Suppose $f: X \to Y$ and $g: Y \to Z$ are functions and $g \circ f: X \to Z$ is one-to-one.

   To show that f is one-to-one, suppose x_1 and x_2 are in X and $f(x_1) = f(x_2)$. *[We must show that $x_1 = x_2$.]*

   Then $g(f(x_1)) = g(f(x_2))$, and so $(g \circ f)(x_1) = (g \circ f)(x_2)$.

   But $g \circ f$ is one-to-one. Hence $x_1 = x_2$ *[as was to be shown]*.

24. $g \circ f: \mathbf{R} \to \mathbf{R}$ is defined by $(g \circ f)(x) = g(f(x)) = g(x + 3) = -(x + 3)$ for all $x \in \mathbf{R}$.

   Since $z = -(x + 3)$ if, and only if, $x = -z - 3$, $(g \circ f)^{-1}: \mathbf{R} \to \mathbf{R}$ is defined by $(g \circ f)^{-1}(z) = -z - 3$ for all $z \in \mathbf{R}$.

   Since $z = -y$ if, and only if, $y = -z$, $g^{-1}: \mathbf{R} \to \mathbf{R}$ is defined by $g^{-1}(z) = -z$ for all $z \in \mathbf{R}$.

   Since $y = x + 3$ if, and only if, $x = y - 3$, $f^{-1}: \mathbf{R} \to \mathbf{R}$ is defined by $f^{-1}(y) = y - 3$.

   $f^{-1} \circ g^{-1}: \mathbf{R} \to \mathbf{R}$ is defined by $(f^{-1} \circ g^{-1})(z) = f^{-1}(g^{-1}(z)) = f^{-1}(-z) = (-z) - 3 = -z - 3$ for all $z \in \mathbf{R}$.

   By the above and the definition of equality of functions, $(g \circ f)^{-1} = f^{-1} \circ g^{-1}$.

27. The property is true.

    <u>Proof 1</u>: Let X, Y, and Z be any sets, let $f : X \to Y$ and $g : Y \to Z$ be any functions, and let C be any subset of Z.

    Proof that $((g \circ f)^{-1}(C) \subseteq f^{-1}(g^{-1}(C))$:

    Suppose $x \in (g \circ f)^{-1}(C)$. *[We must show that $x \in f^{-1}(g^{-1}(C))$.]*

    By definition of inverse image (for $g \circ f$), $(g \circ f)(x) \in C$, and so,

    by definition of composition of functions, $g(f(x)) \in C$.

    Then by definition of inverse image (for g), $f(x) \in g^{-1}(C)$, and

    by definition of inverse image (for f), $x \in f^{-1}(g^{-1}(C))$.

    So by definition of subset, $(g \circ f)^{-1}(C) \subseteq f^{-1}(g^{-1}(C))$.

    Proof that $f^{-1}(g^{-1}(C)) \subseteq (g \circ f)^{-1}(C)$:

    Suppose $x \in f^{-1}(g^{-1}(C))$. *[We must show that $x \in (g \circ f)^{-1}(C)$.]*

    By definition of inverse image (for f), $f(x) \in g^{-1}(C)$, and so,

    by definition of inverse image (for g), $g(f(x)) \in C$.

    So by definition of composition of functions, $(g \circ f)(x) \in C$.

    Then by definition of inverse image (for $g \circ f$), $x \in (g \circ f)^{-1}(C)$.

    So by definition of subset, $f^{-1}(g^{-1}(C)) \subseteq (g \circ f)^{-1}(C)$.

    Conclusion: Since each set is a subset of the other, the two sets are equal.

    <u>Proof 2</u> *(using the logic of if-and-only-if statements)*

    Let X, Y, and Z be any sets, let $f : X \to Y$ and $g : Y \to Z$ be any functions, and let C be any subset of Z.

    Then $x \in (g \circ f)^{-1}(C)$

    $\Leftrightarrow (g \circ f)(x) \in C$ *[by definition of inverse image for $g \circ f$]*

    $\Leftrightarrow g(f(x)) \in C$ *[by definition of composition of functions]*

    $\Leftrightarrow f(x) \in g^{-1}(C)$ *[by definition of inverse image for g]*

    $\Leftrightarrow x \in f^{-1}(g^{-1}(C))$ *[by definition of inverse image for f]*.

    So both sets consist of the same elements, and thus, by definition of set equality, $(g \circ f)^{-1}(C) = f^{-1}(g^{-1}(C))$.

Section 7.4

6. ***Part 1***: The function $I : 2\mathbf{Z} \to \mathbf{Z}$ is defined as follows: $I(n) = n$ for all even integers n. I is clearly one-to-one because if $I(n_1) = I(n_2)$ then by definition of I, $n_1 = n_2$. But I is not onto because the range of I consists only of even integers. In other words, if m is any odd integer, then $I(n) \neq m$ for any even integer n.

   Part 2: The function $J : \mathbf{Z} \to 2\mathbf{Z}$ is defined as follows $J(n) = 2 \lfloor n/2 \rfloor$ for all integers n. Then J is onto because for any even integer m, $m = 2k$ for some integer k. Let $n = 2k$. Then $J(n) = J(2k) = 2 \lfloor 2k/2 \rfloor = 2 \lfloor k \rfloor = 2k = m$. But J is not one-to-one because, for example, $J(2) = 2 \lfloor 2/2 \rfloor = 2 \cdot 1 = 2$ and $J(3) = 2 \lfloor 3/2 \rfloor = 2 \cdot 1 = 2$, so $J(2) = J(3)$ but $2 \neq 3$.

   (More generally, given any integer k, if $m = 2k$, then $J(m) = 2 \lfloor m/2 \rfloor = 2 \lfloor 2k/2 \rfloor = 2 \lfloor k \rfloor = J(m)$ and $J(m+1) = 2 \lfloor (m+1)/2 \rfloor = 2 \lfloor (2k+1)/2 \rfloor = 2 \lfloor k + 1/2 \rfloor = 2k$. So $J(m) = J(m+1)$ but $m \neq m + 1$.)

9. Proof:

Define a function $f: \mathbf{Z}^+ \to \mathbf{Z}^{nonneg}$ as follows: $f(n) = n - 1$ for all positive integers n.

Observe that if $n \geq 1$ then $n - 1 \geq 0$, so f is well-defined.

In addition, f is one-to-one because for all positive integers n_1 and n_2, if $f(n_1) = f(n_2)$ then $n_1 - 1 = n_2 - 1$ and hence $n_1 = n_2$.

Moreover f is onto because if m is any nonnegative integer, then $m + 1$ is a positive integer and $f(m + 1) = (m + 1) - 1 = m$ by definition of f.

Thus, because there is a function $f: \mathbf{Z}^+ \to \mathbf{Z}^{nonneg}$ that is one-to-one and onto, $\mathbf{Z}^+$ has the same cardinality as $\mathbf{Z}^{nonneg}$.

It follows that $\mathbf{Z}^{nonneg}$ is countably infinite and hence countable.

12. Proof:

Define $F: S \to W$ by the rule $F(x) = (b - a)x + a$ for all real numbers x in S.

Then F is well-defined because if $0 < x < 1$, then $a < (b - a)x + a < b$.

In addition, F is one-to-one because if x_1 and x_2 are in S and $F(x_1) = F(x_2)$, then $(b - a)x_1 + a = (b - a)x_2 + a$ and so *[by subtracting a and dividing by $b - a$]* $x_1 = x_2$.

Furthermore, F is onto because if y is any element in W, then $a < y < b$ and so $0 < (y - a)/(b - a) < 1$.

Consequently, $(y - a)/(b - a) \in S$ and $h((y - a)/(b - a)) = (b - a)[(y - a)/(b - a)] + a = y$.

Hence F is a one-to-one correspondence, and so S and W have the same cardinality.

15. Proof:

Let B be the set of all bit strings (strings of 0's and 1's).

Define a function $F: \mathbf{Z}^+ \to B$ as follows: $F(1) = \epsilon$, $F(2) = 0$, $F(3) = 1$, $F(4) = 00$, $F(5) = 01$, $F(6) = 10$, $F(7) = 11$, $F(8) = 000$, $F(9) = 001$, $F(10) = 010$, and so forth.

At each stage, all the strings of length k are counted before the strings of length $k + 1$, and the strings of length k are counted in order of increasing magnitude when interpreted as binary representations of integers.

Thus the set of all bit strings is countably infinite and hence countable.

Note: A more formal definition for F is the following:

$$F(n) = \begin{cases} \epsilon & \text{if } n = 1 \\ \text{the } k\text{-bit binary representation of } n - 2^k & \text{if } n > 1 \text{ and } \lfloor \log_2 n \rfloor = k. \end{cases}$$

For example, $F(7) = 11$ because $\lfloor \log_2 7 \rfloor = 2$ and the two-bit binary representation of $7 - 2^2$ $(= 3)$ is 11.

18. No. For instance, both $\sqrt{2}$ and $-\sqrt{2}$ are irrational (by Theorem 4.7.1 and exercise 23 in Section 4.6), and yet their average is $(\sqrt{2} + (-\sqrt{2}))/2$ which equals 0 and is rational.

More generally: If r is any rational number and x is any irrational number, then both $r + x$ and $r - x$ are irrational (by the result of exercise 12 in Section 4.6 or by the combination of Theorem 4.6.3 and exercise 9 in Section 4.6). Yet the average of these numbers is $((r+x)+(r-x))/2 = r$, which is rational.

21. *Two examples of many:* Define $F: \mathbf{Z} \to \mathbf{Z}$ by the rule $F(n) = \begin{cases} n/2 & \text{if } n \text{ is even} \\ 0 & \text{if } n \text{ is odd} \end{cases}$. Then F is onto because given any integer m, $m = F(2m)$. But F is not one-to-one because, for instance, $F(1) = F(3) = 0$.

Define $G: \mathbf{Z} \to \mathbf{Z}$ by the rule $G(n) = \lfloor n/2 \rfloor$ for all integers n. Then G is onto because given any integer m, $m = \lfloor m \rfloor = \lfloor (2m)/2 \rfloor = G(2m)$. But G is not one-to-one because, for instance, $G(2) = \lfloor 2/2 \rfloor = 1$ and $G(3) = \lfloor 3/2 \rfloor = 1$ and $2 \neq 3$.

24. The proof given below is adapted from one in *Foundations of Modern Analysis* by Jean Dieudonné, New York: Academic Press, 1969, page 14.

    <u>Proof</u>: : Suppose (a, b) and (c, d) are in $\mathbf{Z}^+ \times \mathbf{Z}^+$ and $(a, b) \neq (c, d)$.

    Case 1, $a + b \neq c + d$: By interchanging (a, b) and (c, d) if necessary, we may assume that $a + b < c + d$. Then

$$H(a, b) \quad = \quad b + \frac{(a + b)(a + b + 1)}{2} \qquad \text{by definition of } H$$

$$\Rightarrow \quad H(a, b) \quad \leq \quad a + b + \frac{(a + b)(a + b + 1)}{2} \qquad \text{because } a \geq 0$$

$$\Rightarrow \quad H(a, b) \quad < \quad (a + b + 1) + \frac{(a + b)(a + b + 1)}{2} \qquad \text{because } a + b < a + b + 1$$

$$\Rightarrow \quad H(a, b) \quad < \quad \frac{2(a + b + 1)}{2} + \frac{(a + b)(a + b + 1)}{2}$$

$$\Rightarrow \quad H(a, b) \quad < \quad \frac{(a + b + 1)(a + b + 2)}{2} \qquad \text{by factoring out } (a + b + 1)$$

$$\Rightarrow \quad H(a, b) \quad < \quad \frac{(c + d)(c + d + 1)}{2} \qquad \begin{array}{l}\text{since } a + b < c + d \text{ and } a, b, c, \\ \text{and } d \text{ are integers, } a + b + 1 \leq c + d\end{array}$$

$$\Rightarrow \quad H(a, b) \quad < \quad d + \frac{(c + d)(c + d + 1)}{2} \qquad \text{because } d \geq 0$$

$$\Rightarrow \quad H(a, b) \quad < \quad H(c, d) \qquad \text{by definition of } H.$$

    Therefore, $H(a, b) \neq H(c, d)$.

    Case 2, $a + b = c + d$: First observe that in this case $b \neq d$. For if $b = d$, then subtracting b from both sides of $a + b = c + d$ gives $a = c$, and so $(a, b) = (c, d)$, which contradicts our assumption that $(a, b) \neq (c, d)$. Hence,

$$H(a, b) = b + \frac{(a + b)(a + b + 1)}{2} = b + \frac{(c + d)(c + d + 1)}{2} \neq d + \frac{(c + d)(c + d + 1)}{2} = H(c, d),$$

    and so $H(a, b) \neq H(c, d)$.

    Thus both in case 1 and in case 2, $H(a, b) \neq H(c, d)$, and hence H is one-to-one.

30. <u>Proof by contradiction</u>:

    Suppose not. That is, suppose the set of all irrational numbers were countable.

    Then the set of all real numbers could be written as a union of two countably infinite sets: the set of all rational numbers and the set of all irrational numbers.

    By exercise 29 this union is countably infinite, and so the set of all real numbers would be countably infinite and hence countable.

    But this contradicts the fact that the set of all real numbers is uncountable (which follows immediately from Theorems 7.4.2 and 7.4.3 or Corollary 7.4.4).

    Hence the set of all irrational number is uncountable.

33. <u>Proof</u>:

    First note that there are as many equations of the form $x^2 + bx + c = 0$ as there are pairs (b, c) where b and c are in $\mathbf{Z}$.

    By exercise 32, the set of all such pairs is countably infinite, and so the set of equations of the form $x^2 + bx + c = 0$ is countably infinite.

    Next observe that, by the quadratic formula, each equation $x^2 + bx + c = 0$ has at most two solutions (which may be complex numbers):

$$x = \frac{-b + \sqrt{b^2 - 4c}}{2} \quad \text{and} \quad x = \frac{-b - \sqrt{b^2 - 4c}}{2}.$$

Let

$$R_1 = \left\{ x \mid x = \frac{-b + \sqrt{b^2 - 4c}}{2} \quad \text{for some integers } b \text{ and } c \right\},$$

$$R_2 = \left\{ x \mid x = \frac{-b - \sqrt{b^2 - 4c}}{2} \quad \text{for some integers } b \text{ and } c \right\},$$

and $R = R_1 \cup R_2$. Then R is the set of all solutions of equations of the form $x^2 + bx + c = 0$ where b and c are integers.

Define functions F_1 and F_2 from the set of equations of the form $x^2 + bx + c = 0$ to the sets R_1 and R_2 as follows:

$$F_1(x^2 + bx + c = 0) = \frac{-b + \sqrt{b^2 - 4c}}{2} \quad \text{and} \quad F_2(x^2 + bx + c = 0) = \frac{-b - \sqrt{b^2 - 4c}}{2}.$$

Then F_1 and F_2 are onto functions defined on countably infinite sets, and so, by exercise 27, R_1 and R_2 are countable. Since any union of two countable sets is countable (exercise 31), $R = R_1 \cup R_2$ is countable.

36. <u>Proof</u>:

Let B be the set of all functions from $\mathbf{Z}^+$ to $\{0, 1\}$ and let D be the set of all functions from $\mathbf{Z}^+$ to $\{0, 1, 2, 3, 4, 5, 6, 7, 8, 9\}$.

Elements of B can be represented as infinite sequences of 0's and 1's (for instance, 01101010110...) and elements of D can be represented as infinite sequences of digits from 0 to 9 inclusive (for instance, 20775931124...).

We define a function $H: B \to D$ as follows: For each function f in B, consider the representation of f as an infinite sequence of 0's and 1's.

Such a sequence is also an infinite sequence of digits chosen from 0 to 9 inclusive (one formed without using 2,3,...,9), which represents a function in D. We define this function to be $H(f)$.

More formally, for each $f \in B$, let $H(f)$ be the function in D defined by the rule $H(f)(n) = f(n)$ for all $n \in \mathbf{Z}^+$.

It is clear from the definition that H is one-to-one.

We define a function $K: D \to B$ as follows: For each function g in D, consider the representation of g as a sequence of digits from 0 to 9 inclusive.

Replace each of these digits by its 4-bit binary representation adding leading 0's if necessary to make a full four bits. (For instance, 2 would be replaced by 0010.)

The result is an infinite sequence of 0's and 1's, which represents a function in B. This function is defined to be $K(g)$.

Note that K is one-to-one because if $g_1 \neq g_2$ then the sequences representing g_1 and g_2 must have different digits in some position m, and so the corresponding sequences of 0's and 1's will differ in at least one of the positions $4m - 3, 4m - 2, 4m - 1,$ or $4m$, which are the locations of the 4-bit binary representations of the digits in position m.

It can be shown that whenever there are one-to-one functions from one set to a second and from the second set back to the first, then the two sets have the same cardinality. This fact is known as the Schröder-Bernstein theorem after its two discoverers. For a proof see, for example, *Set Theory and Metric Spaces* by Irving Kaplansky, *A Survey of Modern Algebra*, Third Edition, by Garrett Birkhoff and Saunders MacLane, *Naive Set Theory* by Paul Halmos, or *Topology* by James R. Munkres. The above discussion shows that there are one-to-one functions from B to D and from D to B, and hence by the Schröder-Bernstein theorem the two sets have the same cardinality.

Review Guide: Chapter 7

Definitions: How are the following terms defined?

- function f from a set X to a set Y *(p. 384)*
- Let f be a function from a set X to a set Y.
  - the domain, co-domain, and range of f *(p. 384)*
  - the value of f at x, where x is in X *(p. 384)*
  - the image of x under f, where x is in X *(p. 384)*
  - the output of f for the input x, where x is in X *(p. 384)*
  - the image of X under f *(p. 384)*
  - an inverse image of y, where y is in Y *(p. 384)*
  - the identity function on a set *(p. 387)*
  - the image of A, where $A \subseteq X$ *(p. 392)*
  - the inverse image of B, where $B \subseteq Y$ *(p. 392)*
- logarithm with base b of a positive number x and the logarithmic function with base b *(p. 388)*
- Hamming distance function *(p. 389)*
- Boolean function *(p. 390)*
- one-to-one function *(p. 397)*
- onto function *(p. 402)*
- exponential function with base b *(p. 405)*
- one-to-one correspondence *(p. 408)*
- inverse function *(p. 411)*
- composition of functions *(p. 417)*
- cardinality *(pp. 428-429)*
- countable and uncountable sets. *(p. 431)*

General Function Facts

- How do you draw an arrow diagram for a function defined on a finite set? *(p. 384)*
- Given a function defined by an arrow diagram or by a formula, how do you find values of the function, the range of the function, and the inverse image of an element in its co-domain? *(p. 385)*
- How do you show that two functions are equal? *(p. 386)*
- What is the relation between a sequence and a function? *(p. 387)*
- Can you give an example of a function defined on a power set? a function defined on a Cartesian product? *(p. 387-388)*
- What is an example of an encoding function? a decoding function? *(p. 389)*
- If the claim is made that a given formula defines a function from a set X to a set Y, how do you determine that the "function" is not well-defined? *(p. 391)*

One-to-one and Onto

- How do you show that a function is not one-to-one? *(pp. 397-400)*
- How do you show that a function defined on an infinite set is one-to-one? *(pp. 399-400)*
- How do you show that a function is not onto? *(pp. 402-405)*
- How do you show that a function defined on an infinite set is onto? *(pp. 403-405)*
- How do you determine if a given function has an inverse function? *(p. 411)*
- How do you find an inverse function if it exists? *(pp. 411-413)*

Exponents and Logarithms

- What are the four laws of exponents? *(p. 406)*
- What are the properties of logarithms that correspond to the laws of exponents? *(p. 406)*
- How can you use the laws of exponents to derive properties of logarithms? *(p. 407)*
- How are the logarithmic function with base b and the exponential function with base b related? *(p. 411)*

Composition of Functions

- How do you compute the composition of two functions? *(pp. 417-419)*
- What is the composition of a function with its inverse? *(p. 421)*
- Why is a composition of one-to-one functions one-to-one? *(pp. 421-422)*
- Why is a composition of onto functions onto? *(pp. 423-424)*

Applications of Functions

- What is a Hash function? *(p. 401)*
- How do you show that one set has the same cardinality as another? *(pp. 429-430)*
- How do you show that a given set is countably infinite? countable? *(p. 431)*
- How do you show that the set of all positive rational numbers is countable? *(p. 433)*
- How is the Cantor diagonalization process used to show that the set of real numbers between 0 and 1 is uncountable? *(pp. 433-435)*
- How do you show that the set of all computer programs in a given computer language is countable? *(pp. 437-438)*

Chapter 8: Relations

The first section of this chapter focuses on understanding equivalent ways to specify and represent relations, both finite and infinite. In Section 8.2 the reflexivity, symmetry, and transitivity properties of binary relations are introduced and explored, and in Section 8.3 equivalence relations are discussed. As you work on these sections, you will frequently use the fact that the same proof outlines are used to prove and disprove universal conditional statements no matter what their mathematical context.

Section 8.4 deepens and extends the discussion of congruence relations in Sections 8.2 and 8.3 through applications to modular arithmetic and cryptography. The section is designed to make it possible to give you meaningful practice with RSA cryptography without having to spend several weeks on the topic. After a brief introduction to the idea of cryptography, the first part of the section is devoted to helping you develop the facility with modular arithmetic that is needed to perform the computations for RSA cryptography, especially finding least positive residues of integers raised to large positive powers and using the Euclidean algorithm to compute positive inverses modulo a number. Proofs of the underlying mathematical theory are left to the end of the section.

Section 8.5 introduces another type of binary relation that is especially important in computer science: partial order relations

Section 8.1

3. *c. One possible answer*: 4, 7, 10, -2, -5

   d. One possible answer: 5, 8, 11, -1, -4

   e. <u>Theorem</u>:

   1. All integers of the form $3k$ are related by T to 0.
   2. All integers of the form $3k + 1$ are related by T to 1.
   3. All integers of the form $3k + 2$ are related by T to 2.

   Proof of (2): Let n be any integer of the form $n = 3k + 1$ for some integer k. By substitution, $n - 1 = (3k + 1) - 1 = 3k$, and so by definition of divisibility, $3 \mid (n - 1)$. Hence by definition of T, $n \, T \, 1$.

   The proofs of (1) and (3) are identical to the proof of (2) with 0 and 2, respectively, substituted in place of 1.

6. *b.* Yes, because $\{a, b\} \cap \{b, c\} = \{b\} \neq \emptyset$.   *c.* Yes, because $\{a, b\} \cap \{a, b, c\} = \{a, b\} \neq \emptyset$.

9. *c.* No, because the sum of the characters in 2212 is 7 and the sum of the characters in 2121 is 6, and $7 \neq 6$.

   d. Yes, because the sum of the characters in 1220 is 5 and the sum of the characters in 2111 is 5, and $5 = 5$.

12. *b.* No. If $F: X \to Y$ is not one-to-one, then there exist x_1 and x_2 in X and y in Y such that $(x_1, y) \in F$ and $(x_2, y) \in F$ and $x_1 \neq x_2$. But this implies that there exist x_1 and x_2 in X and y in Y such that $(y, x_1) \in F^{-1}$ and $(y, x_2) \in F^{-1}$ and $x_1 \neq x_2$. Consequently, F^{-1} does not satisfy property (2) of the definition of function.

112

18.

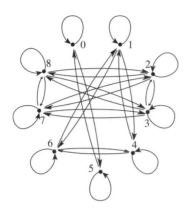

24. *b.*   466581   Mary Lazars
            778400   Jamal Baskers

Section 8.2

18. **Q is reflexive**: Suppose x is any real number. *[We must show that x Q x.]* By definition of Q, this means that $x - x$ is rational. But this is true because $x - x = 0$, and 0 is rational since $0 = 0/1$. *[So x Q x as was to be shown.]*

    Q is symmetric: Suppose x and y are any real numbers such that x Q y. *[We must show that y Q x.]* By definition of Q, $x - y$ is rational. Now $y - x = -(x - y)$ and the negative of any rational number is rational *[by exercise 13, Section 4.2]*. Hence $y - x$ is rational, and so y Q x by definition of Q *[as was to be shown]*.

    Q is transitive: Suppose x, y and z are any real numbers such that x Q y and y Q z. *[We must show that x Q z.]* By definition of Q, $x - y$ is rational and $y - z$ is rational.

    Then, since $x - p = (x - y) + (y - z)$, we have that $x - z$ is a sum of rational numbers, and hence $x - z$ is rational *[by Theorem 4.2.2]*. Thus, by definition of Q, x Q z *[as was to be shown]*.

21. For each set X, let $N(X)$ be the number of elements in X.

    L *is not reflexive*: **L** is reflexive $\Leftrightarrow$ for all sets A $\in$ $\mathscr{P}(X)$, A **L** A. By definition of **L** this means that for all sets A in $\mathscr{P}(X)$, $N(A) < N(A)$. But this is false for every set in $\mathscr{P}(X)$. For instance, let A $= \emptyset$. Then $N(A) = 0$, and 0 is not less than 0.

    L *is not symmetric*: For **L** to be symmetric would mean that for all sets A and B in $\mathscr{P}(X)$, if A **L** B then B **L** A. By definition of **L**, this would mean that for all sets A and B in $\mathscr{P}(X)$, if $N(A) < N(B)$, then $N(B) < N(A)$. But this is false for all sets A and B in $\mathscr{P}(X)$. For instance, take A $= \emptyset$ and B $= \{a\}$. Then $N(A) = 0$ and $N(B) = 1$. It follows that A is related to B by **L** (since $0 < 1$), but B is not related to A by **L** (since $1 \not< 0$).

    L *is transitive*: To prove transitivity of **L**, we must show that for all sets A, B, and C in $\mathscr{P}(X)$, if A **L** B and B **L** C then A **L** C. By definition of **L** this means that for all sets A, B, and C in $\mathscr{P}(X)$, if $N(A) < N(B)$ and $N(B) < N(C)$, then $N(A) < N(C)$. But this is true by the transitivity property of order (Appendix A, T18).

24. **U *is not reflexive***: **U** is reflexive $\Leftrightarrow$ for all sets A in $\mathscr{P}(X)$, A **U** A. By definition of **U** this means that for all sets A in $\mathscr{P}(X)$, A $\neq$ A. But this is false for every set in $\mathscr{P}(X)$. For instance, let A $= \emptyset$. It is not true that $\emptyset \neq \emptyset$.

    U *is symmetric*: **U** is symmetric $\Leftrightarrow$ for all sets A and B in $\mathscr{P}(X)$, if A **U** B then B **U** A. By definition of **U**, this means that for all sets A and B in $\mathscr{P}(X)$, if A $\neq$ B, then B $\neq$ A. But this is true.

U *is not transitive*: **U** is transitive $\Leftrightarrow$ for all sets A, B, and C in $\mathscr{P}(X)$, if A **U** B and B **U** C then A **U** C. By definition of **U** this means that for all sets A, B, and C in $\mathscr{P}(X)$, if A $\neq$ B and B $\neq$ Z, then A $\neq$ C. But this is false as the following counterexample shows. Since $X \neq \emptyset$, there exists an element x in X. Let A $= \{x\}$, B $= \emptyset$, and C $= \{x\}$. Then A $\neq$ B and B $\neq$ Z, but A $=$ C.

30. **R** *is reflexive*: R is reflexive $\Leftrightarrow$ for all points p in A, $p\,R\,p$. By definition of R this means that for all elements p in A, p and p both lie on the same half line emanating from the origin. But this is true.

    R *is symmetric*:: *[We must show that for all points p_1 and p_2 in A, if $p_1 R\, p_2$ then $p_2 R\, p_1$.]* Suppose p_1 and p_2 are points in A such that $p_1 R\, p_2$. By definition of R this means that p_1 and p_2 lie on the same half line emanating from the origin. But this implies that p_2 and p_1 lie on the same half line emanating from the origin. So by definition of R, $p_2 R\, p_1$.

    R *is transitive*: *[We must show that for all points p_1, p_2 and p_3 in A, if $p_1 R\, p_2$ and $p_2 R\, p_3$ then $p_1 R\, p_3$.]* Suppose p_1, p_2, and p_3 are points in A such that $p_1 R\, p_2$ and $p_2 R\, p_3$. By definition of R, this means that p_1 and p_2 lie on the same half line emanating from the origin and p_2 and p_3 lie on the same half line emanating from the origin.

    Since two points determine a line, it follows that both p_1 and p_3 lie on the same half line determined by the origin and p_2. Thus p_1 and p_3 lie on the same half line emanating from the origin. So by definition of R, $p_1 R\, p_3$.

33. **R** *is not reflexive*: R is reflexive $\Leftrightarrow$ for all lines l in A, $l\,R\,l$. By definition of R this means that for all lines l in the plane, l is perpendicular to itself. But this is false for every line in the plane.

    R *is symmetric*: *[We must show that for all lines l_1 and l_2 in A, if $l_1 R\, l_2$ then $l_2 R\, l_1$.]* Suppose l_1 and l_2 are lines in A such that $l_1 R\, l_2$. By definition of R this means that l_1 is perpendicular to l_2. But this implies that l_2 is perpendicular to l_1. So by definition of R, $l_2 R\, l_1$.

    R *is not transitive*: R is transitive $\Leftrightarrow$ for all lines l_1, l_2, and l_3 in A, if $l_1 R\, l_2$ and $l_2 R\, l_3$ then $l_1 R\, l_3$. But this is false. As a counterexample, take l_1 and l_3 to be the horizontal axis and l_2 to be the vertical axis. Then $l_1 R\, l_2$ and $l_2 R\, l_3$ because the horizontal axis is perpendicular to the vertical axis and the vertical axis is perpendicular to the horizontal axis. But $l_1 \not{R}\, l_3$ because the horizontal axis is not perpendicular to itself.

36. The statement is true.

    <u>Proof</u>: Suppose R is a transitive relation on a set A. To show that R^{-1} is transitive, we suppose that x, y, and z are any elements of A such that $x\,R^{-1}\,y$ and $y\,R^{-1}\,z$. *[We must show that $x\,R^{-1}\,z$.]* By definition of R^{-1}, $y\,R\,x$ and $z\,R\,y$, or, equivalently, $z\,R\,y$ and $y\,R\,x$. Since R is transitive, $z\,R\,x$. Thus, by definition of R^{-1}, $z\,R^{-1}\,x$ *[as was to be shown]*.

39. **R $\cap$ S** *is transitive*: Suppose R and S are transitive. *[To show that $R \cap S$ is transitive, we must show that $\forall x, y, z \in A$, if $(x, y) \in R \cap S$ and $(y, z) \in R \cap S$ then $(x, z) \in R \cap S$.]* So suppose x, y, and z are elements of A such that $(x, y) \in R \cap S$ and $(y, z) \in R \cap S$. By definition of intersection, $(x, y) \in R$, $(x, y) \in S$, $(y, z) \in R$, and $(y, z) \in S$. It follows that $(x, z) \in R$ because R is transitive and $(x, y) \in R$ and $(y, z) \in R$. Also $(x, z) \in S$ because S is transitive and $(x, y) \in S$ and $(y, z) \in S$. Thus by definition of intersection $(x, z) \in R \cap S$.

42. **R $\cup$ S** *is not necessarily transitive*: As a counterexample, let $R = \{(0, 1)\}$ and $S = \{(1, 2)\}$. Then both R and S are transitive (by default), but $R \cup S = \{(0, 1), (1, 2)\}$ is not transitive because $(0, 1) \in R \cup S$ and $(1, 2) \in R \cup S$ but $(0, 2) \notin R \cup S$. As another counterexample, let $R = \{(x, y) \in \mathbf{R} \times \mathbf{R} \mid x < y\}$ and let $S = \{(x, y) \in \mathbf{R} \times \mathbf{R} \mid x > y\}$.

    Then both R and S are transitive because of the transitivity of order for the real numbers. But $R \cup S = \{(x, y) \in \mathbf{R} \times \mathbf{R} \mid x \neq y\}$ is not transitive because, for instance, $(1, 2) \in R \cup S$ and $(2, 1) \in R \cup S$ but $(1, 1) \notin R \cup S$.

Section 8.3

6. distinct equivalence classes: $\{0, 3, -3\}, \{1, 4, -2\}, \{2, 5, -1, -4\}$

9. distinct equivalence classes: $\{\emptyset, \{0\}, \{1, -1\}, \{-1, 0, 1\}\}, \{\{1\}, \{0, 1\}\}, \{\{-1\}, \{0, -1\}\}$

12. $[0] = \{x \in A \mid 5 \text{ divides } (x^2 - 0)\} = \{0\}$

    $[1] = \{x \in A \mid 5 \text{ divides } (x^2 - 1)\} = \{x \in A \mid 5 \text{ divides } x - 1)(x + 1)\} = \{1, -1, 4, -4\}$

    $[2] = \{x \in A \mid 5 \text{ divides } (x^2 - 2^2)\} = \{x \in A \mid 5 \text{ divides } (x - 2)(x + 2)\} = \{2, -2, 3, -3\}$

15. *b.* false   *c.* true   *d.* true

18. *b.* Let $A_1 = \{1, 2\}$, $A_2 = \{2, 3\}$, $x = 1$, $y = 2$, and $z = 3$. Then both x and y are in A_1 and both y and z are in A_2, but x and z are not both in either A_1 or A_2.

21. (1) <u>Proof</u>:

    F is reflexive: Suppose m is any integer. Since $m - m = 0$ and $4 \mid 0$, we have that $4 \mid (m - m)$. Consequently, $m \, F \, m$ by definition of F.

    F is symmetric: Suppose m and n are any integers such that $m \, F \, n$. By definition of F this means that $4 \mid (m - n)$, and so, by definition of divisibility, $m - n = 4k$ for some integer k. Now $n - m = -(m - n)$. Hence by substitution, $n - m = -(4k) = 4 \cdot (-k)$. It follows that $4 \mid n - m$ by definition of divisibility (since $-k$ is an integer), and thus $n \, F \, m$ by definition of F.

    F is transitive: Suppose m, n and p are any integers such that $m \, F \, n$ and $n \, F \, p$. By definition of F, this means that $4 \mid (m - n)$ and $4 \mid (n - p)$, and so, by definition of divisibility, $m - n = 4k$ for some integer k, and $n - p = 4l$ for some integer l. Now $m - p = (m - n) + (n - p)$. Hence by substitution, $m - p = 4k + 4l = 4(k + l)$. It follows that $4 \mid (m - p)$ by definition of divisibility (since $k + l$ is an integer), and thus $m \, F \, p$ by definition of F.

    F is an equivalence relation because it is reflexive, symmetric, and transitive.

    (2) Four distinct classes: $\{x \in \mathbf{Z} \mid x = 4k \text{ for some integer } k\}$, $\{x \in \mathbf{Z} \mid x = 4k + 1 \text{ for some integer } k\}$, $\{x \in \mathbf{Z} \mid x = 4k + 2 \text{ for some integer } k\}$, $\{x \in \mathbf{Z} \mid x = 4k + 3 \text{ for some integer } k\}$

24. (1) <u>Proof</u>:

    R is reflexive because for each identifier x in A, x has the same memory location as x.

    R is symmetric because for all identifiers x and y in A, if x has the same memory location as y then y has the same memory location as x.

    R is transitive because for all identifiers x, y, and z in A, if x has the same memory location as y and y has the same memory location as z then x has the same memory location as z.

    R is an equivalence relation because it is reflexive, symmetric, and transitive.

    (2) There are as many distinct equivalence classes as there are distinct memory locations that are used to store variables during execution of the program. Each equivalence class consists of all variables that are stored in the same location.

27. (1) <u>Proof</u>:

    R is reflexive: Suppose m is any integer. Since $m^2 - m^2 = 0$ and $4 \mid 0$, we have that $4 \mid (m^2 - m^2)$. Consequently, $m \, R \, m$ by definition of R.

    R is symmetric: Suppose m and n are any integers such that $m \, R \, n$. By definition of R this means that $4 \mid (m^2 - n^2)$, and so, by definition of divisibility, $m^2 - n^2 = 4k$ for some integer k. Now $n^2 - m^2 = -(m^2 - n^2)$. Hence by substitution, $n^2 - m^2 = -(4k) = 4 \cdot (-k)$. It follows that $4 \mid (n^2 - m^2)$ by definition of divisibility (since $-k$ is an integer), and thus $n \, R \, m$ by definition of R.

R is transitive: Suppose m, n and p are any integers such that $m \, R \, n$ and $n \, R \, p$. By definition of R, this means that $4 \mid (m^2 - n^2)$ and $4 \mid (n^2 - p^2)$, and so, by definition of divisibility, $m^2 - n^2 = 4k$ for some integer k, and $n^2 - p^2 = 4l$ for some integer l. Now $m^2 - p^2 = (m^2 - n^2) + (n^2 - p^2)$. Hence by substitution, $m^2 - p^2 = 4k + 4l = 4(k + l)$. It follows that $4 \mid (m^2 - p^2)$ by definition of divisibility (since $k + l$ is an integer), and thus $m \, R \, p$ by definition of R.

R is an equivalence relation because it is reflexive, symmetric, and transitive.

(2) If m is even, then $m = 2a$ for some integer a, and so $m^2 - 0^2 = (2a)^2 = 4a^2$, which is divisible by 4. Hence $m \in [0]$.

If m is odd, then $m = 2a+1$ for some integer a, and so $m^2 - 1^2 = (2a+1)^2 - 1 = 4a^2 + 4a + 1 - 1 = 4a^2 + 4a$, which is divisible by 4. Hence $m \in [1]$.

Thus there are two distinct equivalence classes:

$$[0] = \{m \in \mathbf{Z} \mid m \text{ is even}\} \text{ and } [1] = \{m \in \mathbf{Z} \mid m \text{ is odd}\}.$$

30. (1) Proof:

Q is reflexive because each ordered pair has the same second element as itself.

Q is symmetric for the following reason: Suppose (w, x) and (y, z) are ordered pairs of real numbers such that $(w, x) \, Q \, (y, z)$. Then, by definition of Q, $x = z$. By the symmetric property of equality, this implies that $z = x$, and so, by definition of Q, $(y, z) \, Q \, (w, x)$.

Q is transitive for the following reason: Suppose (u, v), (w, x), and (y, z) are ordered pairs of real numbers such that $(u, v) \, Q \, (w, x)$ and $(w, x) \, Q \, (y, z)$. Then, by definition of Q, $v = x$ and $x = z$. By the transitive property of equality, this implies that $v = z$, and so, by definition of Q, $(u, v)(y, z) \, Q \, (y, z)$.

Q is an equivalence relation because it is reflexive, symmetric, and transitive.

(2) There is one equivalence class for each real number. The distinct equivalence classes are all the sets of the form $\{(x, y) \in \mathbf{R} \times \mathbf{R} \mid y = b\}$ where b is a real number. Equivalently, the distinct equivalence classes are all the vertical lines in the Cartesian plane.

33. The distinct equivalence classes can be identified with the points on a geometric figure, called a *torus*, that has the shape of the surface of a doughnut.

Each point in the interior of the rectangle $\{(x, y) \mid 0 < x < 1 \text{ and } 0 < y < 1\}$ is only equivalent to itself.

Each point on the top edge of the rectangle is in the same equivalence class as the point vertically below it on the bottom edge of the rectangle (so we can imagine identifying these points by gluing them together — this gives us a cylinder).

In addition, each point on the left edge of the rectangle is in the same equivalence class as the point horizontally across from it on the right edge of the rectangle (so we can also imagine identifying these points by gluing them together — this brings the two ends of the cylinder together to produce a torus).

39. Proof:

Suppose R is an equivalence relation on a set A, a and b are in A, and $[a] = [b]$.

Since R is reflexive, $a \, R \, a$, and so by definition of class, $a \in [a]$. *[Alternatively, one could reference exercise 36 here.]*

Since $[a] = [b]$, by definition of set equality, $a \in [b]$.

But then by definition of equivalence class, $a \, R \, b$.

42. *a.* Suppose $(a, b) \in A$. By commutativity of multiplication for the real numbers, $ab = ba$. But then by definition of R, $(a, b)R(a, b)$, and so R is reflexive.

    b. Suppose $(a, b), (c, d) \in A$ and $(a, b)R(c, d)$. By definition of R, $ad = bc$, and so by commutativity of multiplication for the real numbers and symmetry of equality, $cb = da$. But then by definition of R, $(c, d)R(a, b)$, and so R is symmetric.

    d. For example, (2,5), (4,10), (-2,-5), and (6,15) are all in $[(2,5)]$.

45. The given argument assumes that from the fact that the statement "$\forall x$ in A, if $x \ R \ y$ then $y \ R \ x$" is true, it follows that given any element x in R, there must exist an element y in R such that $x \ R \ y$ and $y \ R \ x$. This is false. For instance, consider the following relation R defined on $A = \{1, 2\} : R = \{(1, 1)\}$. This relation is symmetric and transitive, but it is not reflexive. Given $2 \in A$, there is no element y in A such that $(2, y) \in R$. Thus we cannot go on to use symmetry to say that $(y, 2) \in R$ and transitivity to conclude that $(2, 2) \in R$.

Section 8.4

6. Proof:

    Given any integer $n > 1$ and any integer a with $0 \leq a < n$, the notation $[a]$ denotes the equivalence class of a for the relation of congruence modulo n.

    We first show that given any integer m, m is in one of the classes $[0], [1], [2], \ldots, [n-1]$.

    The reason is that, by the quotient-remainder theorem, $m = nk + a$, where k and a are integers and $0 \leq a < n$, and so, by Theorem 8.4.1, $m \equiv a \pmod{n}$. It follows by Lemma 8.3.2 that $[m] = [a]$.

    Next we use an argument by contradiction to show that all the equivalence classes $[0], [1], [2], \ldots, [n-1]$ are distinct.

    For suppose not. That is, suppose a and b are integers with $0 \leq a < n$ and $0 \leq b < n$, $a \neq b$, and $[a] = [b]$. Without loss of generality, we may assume that $a > b \geq 0$, which implies that $-a < -b \leq 0$. Adding a to all parts of the inequality gives $0 < a - b \leq a$. By Exercise 8.3.39, $[a] = [b]$ implies that $a \equiv b \pmod{n}$. Hence, by Theorem 8.4.1, $n \mid (a - b)$, and so, by Theorem 4.3.1, $n \leq a - b$.. But $a < n$. Thus $n \leq a - b \leq a < n$, which is contradictory. Therefore the supposition is false, and we conclude that all the equivalence classes $[0], [1], [2], \ldots, [n-1]$ are distinct.

9. *b.* Proof:

    Suppose a, b, c, d, and n are integers with $n > 1$, $a \equiv c \pmod{n}$, and $b \equiv d \pmod{n}$. *[We must show that $a - b \equiv (c - d) \pmod{n}$.]*

    By definition, $a - c = nr$ and $b - d = ns$ for some integers r and s. Then

    $$(a - b) - (c - d) = (a - c) - (b - d) = nr - ns = n(r - s).$$

    But $r - s$ is an integer, and so, by definition, $a - b \equiv (c - d) \pmod{n}$.

12. *b.* Proof:

    Suppose a is a positive integer. Then $a = \sum_{k=0}^{n} d_k 10^k$, for some nonnegative integer n and integers d_k where $0 \leq d_k < 10$ for all $k = 1, 2, \ldots, n$. By Theorem 8.4.3,

    $$a = \sum_{k=0}^{n} d_k 10^k \equiv \sum_{k=0}^{n} d_k \cdot 1 \equiv \sum_{k=0}^{n} d_k \pmod{9}$$

    because, by part (a), each $10^k \equiv 1 \pmod{9}$. Hence, by Theorem 8.4.1, both a and $\sum_{k=0}^{n} d_k$ have the same remainder upon division by 9, and thus if either one is divisible by 9, so is the other.

18. $48^1 \ mod \ 713 = 48$

    $48^2 \ mod \ 713 = 165$

    $48^4 \ mod \ 713 = 165^2 \ mod \ 713 = 131$

    $48^8 \ mod \ 713 = 131^2 \ mod \ 713 = 49$

    $48^{16} \ mod \ 713 = 49^2 \ mod \ 713 = 262$

    $48^{32} \ mod \ 713 = 262^2 \ mod \ 713 = 196$

    $48^{64} \ mod \ 713 = 196^2 \ mod \ 713 = 627$

    $48^{128} \ mod \ 713 = 627^2 \ mod \ 713 = 266$

    $48^{256} \ mod \ 713 = 266^2 \ mod \ 713 = 169$

    Hence, by Theorem 8.4.3,

    $$48^{307} = 48^{256+32+16+2+1} = 48^{256}48^{32}48^{16}48^2 48^1 \equiv 169 \cdot 196 \cdot 262 \cdot 165 \cdot 48 \equiv 12 (mod \ 713),$$

    and thus $48^{307} \ mod \ 713 = 12$.

21. The letters in EXCELLENT translate numerically into 05, 24, 03, 05,12, 12, 05, 14, 20. The solutions for exercises 19 (in Appendix B) and 20 (above) show that E, L, and C are encrypted as 15, 23, and 27, respectively. To encrypt X, we compute $24^3 \ mod \ 55 = 19$, to encrypt N, we compute $14^3 \ mod \ 55 = 49$, and to encrypt T, we compute $20^3 \ mod \ 55 = 25$. So the ciphertext is 15 19 27 15 23 23 15 49 25.

24. By Example 8.4.10, the decryption key is 27. Thus the residues modulo 55 for 51^{27}, 14^{27}, 49^{27}, and 15^{27} must be found and then translated into letters of the alphabet. Because $27 = 16 + 8 + 2 + 1$, we first perform the following computations:

$51^1 \equiv 51 \ (mod \ 55)$	$14^1 \equiv 14 \ (mod \ 55)$	$49^1 \equiv 49 \ (mod \ 55)$
$51^2 \equiv 16 \ (mod \ 55)$	$14^2 \equiv 31 \ (mod \ 55)$	$49^2 \equiv 36 \ (mod \ 55)$
$51^4 \equiv 16^2 \equiv 36 \ (mod \ 55)$	$14^4 \equiv 31^2 \equiv 26 \ (mod \ 55)$	$49^4 \equiv 36^2 \equiv 31 \ (mod \ 55)$
$51^8 \equiv 36^2 \equiv 31 \ (mod \ 55)$	$14^8 \equiv 26^2 \equiv 16 \ (mod \ 55)$	$49^8 \equiv 31^2 \equiv 26 \ (mod \ 55)$
$51^{16} \equiv 31^2 \equiv 26 \ (mod \ 55)$	$14^{16} \equiv 16^2 \equiv 36 \ (mod \ 55)$	$49^{16} \equiv 26^2 \equiv 16 \ (mod \ 55)$

    Then

    $51^{27} \ mod \ 55 = (26 \cdot 31 \cdot 16 \cdot 51) \ mod \ 55 = 6,$

    $14^{27} \ mod \ 55 = (36 \cdot 16 \cdot 31 \cdot 14) \ mod \ 55 = 9,$

    $49^{27} \ mod \ 55 = (16 \cdot 26 \cdot 36 \cdot 49) \ mod \ 55 = 14.$

    In addition, we know from the solution to exercise 23 above that $15^{27} \ mod \ 55 = 5$. But 6, 9, 14, and 5 translate into letters as F, I, N, and E. So the message is FINE.

27. *Step 1:* $4158 = 1568 \cdot 2 + 1022$, and so $1022 = 4158 - 1568 \cdot 2$

    Step 2: $1568 = 1022 \cdot 1 + 546$, and so $546 = 1568 - 1022$

    Step 3: $1022 = 546 \cdot 1 + 476$, and so $476 = 1022 - 546$

    Step 4: $546 = 476 \cdot 1 + 70$, and so $70 = 546 - 476$

    Step 5: $476 = 70 \cdot 6 + 56$, and so $56 = 476 - 70 \cdot 6$

    Step 6: $70 = 56 \cdot 1 + 14$, and so $14 = 70 - 56$

    Step 7: $56 = 14 \cdot 4 + 0$, and so $\gcd(4158, 1568) = 14$,

    which is the remainder obtained just before the final division.

    Substitute back through steps 6–1:

    $14 = 70 - 56 = 70 - (476 - 70 \cdot 6) = 70 \cdot 7 - 476$

$$= (546 - 476) \cdot 7 - 476 = 7 \cdot 546 - 8 \cdot 476$$
$$= 7 \cdot 546 - 8 \cdot (1022 - 546) = 15 \cdot 546 - 8 \cdot 1022$$
$$= 15 \cdot (1568 - 1022) - 8 \cdot 1022 = 15 \cdot 1568 - 23 \cdot 1022$$
$$= 15 \cdot 1568 - 23 \cdot (4158 - 1568 \cdot 2) = 61 \cdot 1568 - 23 \cdot 4158$$

(It is always a good idea to verify that no mistake has been made by verifying that the final expression really does equal the greatest common divisor. In this case, a computation shows that the answer is correct.)

30. Proof:

Suppose a and b are positive integers, $S = \{x \mid x \text{ is a positive integer and } x = as + bt \text{ for some integers } s \text{ and } t\}$, and c is the least element of S. We will show that $c \mid b$.

By the quotient-remainder theorem, $b = cq + r$ (*) for some integers q and r with $0 \le r < c$.

Now because c is in S, $c = as + bt$ for some integers s and t. Thus, by substitution into equation (*),

$$r = b - cq = b - (as + bt)q = a(-sq) + b(1 - tq).$$

Hence, by definition of S, either $r = 0$ or $r \in S$.

But if $r \in S$, then $r \ge c$ because c is the least element of S, and thus both $r < c$ and $r \ge c$ would be true, which would be a contradiction.

Therefore, $r \notin S$, and thus by elimination, we conclude that $r = 0$.

It follows that $b - cq = 0$, or, equivalently, $b = cq$, and so $c \mid b$ [as was to be shown].

33. Proof:

Suppose a, b, and c are integers such that $\gcd(a, b) = 1$, $a \mid c$, and $b \mid c$. We will show that $ab \mid c$.

By Corollary 8.4.6 (or by Theorem 8.4.5), there exist integers s and t such that $as + bt = 1$.

Also, by definition of divisibility, $c = au = bv$, for some integers u and v. Hence, by substitution,

$$c = asc + btc = as(bv) + bt(au) = ab(sv + tu).$$

But $sv + tu$ is an integer, and so, by definition of divisibility, $ab \mid c$ [as was to be shown].

42. b. When $a = 8$ and $p = 11$,

$$a^{p-1} = 8^{10} = 1073741824 \equiv 1 (\text{mod } 11) \text{ because } 1073741824 - 1 = 11 \cdot 97612893.$$

Section 8.5

3. R is not antisymmetric.

   Counterexample: Let $s = 0$ and $t = 1$. Then $s \, R \, t$ and $t \, R \, s$ because $l(s) \le l(t)$ and $l(t) \le l(s)$, since both $l(s)$ and $l(t)$ equal 1, but $s \ne t$.

6. R is a partial order relation.

   Proof:

   R is reflexive: Suppose $r \in P$. Then $r = r$, and so by definition of R, $r \, R \, r$.

   R is antisymmetric: Suppose $r, s \in P$ and $r \, R \, s$ and $s \, R \, r$. [We must show that $r = s$.]

   By definition of R, either r is an ancestor of s or $r = s$ and either s is an ancestor of r or $s = r$.

Now it is impossible for both r to be an ancestor of s and s to be an ancestor of r. Hence one of these conditions must be false, and so $r = s$ *[as was to be shown]*.

R is transitive: Suppose $r, s, t \in P$ and $r\ R\ s$ and $s\ R\ t$. *[We must show that $r\ R\ t$.]*

By definition of R, either r is an ancestor of s or $r = s$ and either s is an ancestor of t or $s = t$.

In case r is an ancestor of s and s is an ancestor of t, then r is an ancestor of t, and so $r\ R\ t$.

In case r is an ancestor of s and $s = t$, then r is an ancestor of t, and so $r\ R\ t$.

In case $r = s$ and s is an ancestor of t, then r is an ancestor of t, and so $r\ R\ t$.

In case $r = s$ and $s = t$, then $r = t$, and so $r\ R\ t$. Thus in all four possible cases, $r\ R\ t$ *[as was to be shown]*.

Conclusion: Since R is reflexive, antisymmetric, and transitive, R is a partial order relation.

9. R is not a partial order relation because R is not antisymmetric.

   Counterexample: Let $x = 2$ and $y = -2$. Then $x\ R\ y$ because $(-2)^2 \le 2^2$, and $y\ R\ x$ because $2^2 \le (-2)^2$. But $x \ne y$ because $2 \ne -2$.

12. Proof:

    $\preceq$ is reflexive: Suppose s is in S. If $s = \epsilon$, then $s \preceq s$ by (3). If $s \ne \epsilon$, then $s \preceq s$ by (1). Hence in either case, $s \preceq s$.

    $\preceq$ is antisymmetric: Suppose s and t are in S and $s \preceq t$ and $t \preceq s$. *[We must show that $s = t$.]*

    By definition of S, either $s = \epsilon$ or $s = a_1 a_2 \ldots a_m$ and either $t = \epsilon$ or $t = b_1 b_2 \ldots b_n$ for some positive integers m and n and elements $a_1, a_2, \ldots, a_m$ and $b_1, b_2, \ldots, b_n$ in A.

    It is impossible to have $s \preceq t$ by virtue of condition (2) because in that case there is no circumstance that would give $t \preceq s$.

    [For suppose $s \preceq t$ by virtue of condition (2). Then for some integer k with $k \le m$, $k \le n$, and $k \ge 1$, $a_i = b_i$ for all $i = 1, 2, \ldots, k - 1$, and $a_k\ R\ b_k$ and $a_k \ne b_k$. In this situation, it is clearly impossible for $t \preceq s$ by virtue either of condition (1) or (3), and so, if $t \preceq s$, then it must be by virtue of condition (2). But in that case, since $a_k \ne b_k$, it must follow that $b_k\ R\ a_k$, and so, since R is a partial order relation, $a_k = b_k$. However, this contradicts the fact that $a_k \ne b_k$. Hence it cannot be the case that $s \preceq t$ by virtue of condition (2).]

    Similarly, it is impossible for $t \preceq s$ by virtue of condition (2).

    Hence $s \preceq t$ and $t \preceq s$ by virtue either of condition (1) or of condition (3).

    In case $s \preceq t$ by virtue of condition (1), then neither s nor t is the null string and so $t \preceq s$ by virtue of condition (1). Then by (1) $m \le n$ and $a_i = b_i$ for all $i = 1, 2, \ldots, m$ and $n \le m$ and $b_i = a_i$ for all $i = 1, 2, \ldots, m$, and so, in this case, $s = t$.

    In case $s \preceq t$ by virtue of condition (3), then $s = \epsilon$, and so since $t \preceq s$, $t \preceq \epsilon$. But the only situation that can give this result is condition (3) with $t = \epsilon$. Hence in this case, $s = t = \epsilon$.

    Thus in all possible cases, if $s \preceq t$ and $t \preceq s$, then $s = t$ *[as was to be shown]*.

    $\preceq$ is transitive: Suppose s and t are any elements of S such that $s \preceq t$ and $t \preceq u$. *[We must show that $s \preceq u$.]*

    By definition of S, either $s = \epsilon$ or $s = a_1 a_2 \ldots a_m$, either $t = \epsilon$ or $t = b_1 b_2 \ldots b_n$, and either $u = \epsilon$ or $u = c_1 c_2 \ldots c_p$ for some positive integers m, n, and p and elements $a_1, a_2, \ldots, a_m$, $b_1, b_2, \ldots, b_n$, and $c_1, c_2, \ldots, c_p$ in A.

    Case 1 ($s = \epsilon$): In this case, $s\ R\ u$ by (3).

    Case 2 ($s \ne \epsilon$): In this case, since $s\ R\ t$, $t \ne \epsilon$, and since $t\ R\ u$, $u \ne \epsilon$.

Subcase a (s R t by condition (1) and t R u by condition (1)): Then $m \le n$ and $n \le p$ and $a_i = b_i$ for all $i = 1, 2, \ldots, m$ and $b_j = c_j$ for all $j = 1, 2, \ldots, n$. It follows that $a_i = c_i$ for all $i = i, 2, \ldots, m$, and so by (1), $s\ R\ u$.

Subcase b (s R t by condition (1) and t R u by condition (2)): Then $m \le n$ and $a_i = b_i$ for all $i = 1, 2, \ldots, m$, and for some integer k with $k \le n$, $k \le p$, and $k \ge 1$, $b_j = c_j$ for all $j = 1, 2, \ldots, k - 1$, $b_k\ R\ c_k$, and $b_k \ne c_k$.

If $k \le m$, then s and u satisfy condition (2) *[because $a_i = b_i$ for all $i = 1, 2, \ldots, m$ and so $k \le m$, $k \le p$, $k \ge 1$, $a_i = b_i = c_i$ for all $i = 1, 2, \ldots, k - 1$, $a_k\ R\ c_k$, and $a_k \ne c_k$].*

If $k > m$, then s and u satisfy condition (1) *[because $a_i = b_i = c_i$ for all $i = 1, 2, \ldots, m$].* Thus in either case $s\ R\ u$.

Subcase c (s R t by condition (2) and t R u by condition (1)): Then for some integer k with $k \le m$, $k \le n$, $k \ge 1$, $a_i = b_i$ for all $i = 1, 2, \ldots, k - 1$, $a_k\ R\ b_k$, and $a_k \ne b_k$, and $n \le p$ and $b_j = c_j$ for all $j = i, 2, \ldots, n$. Then s and u satisfy condition (2) *[because $k \le n$, $k \le p$ (since $k \le n$ and $n \le p$), $k \ge 1$, $a_i = b_i = c_i$ for all $i = 1, 2, \ldots, k - 1$ (since $k - 1 < n$), $a_k\ Rc_k$ (since $b_k = c_k$ because $k \le n$), and $a_k \ne c_k$ (since $b_k = c_k$ and $a_k \ne b_k$)].* Thus $s\ R\ u$.

Subcase d (s R t by condition (2) and t R u by condition (2)): Then for some integer k with $k \le m$, $k \le n$, $k \ge 1$, $a_i = b_i$ for all $i = 1, 2, \ldots, k - 1$, $a_k\ R\ b_k$, and $a_k \ne b_k$, and for some integer l with $l \le n$, $l \le p$, and $l \ge 1$, $b_j = c_j$ for all $j = 1, 2, \ldots, l - 1$, $b_l\ R\ c_l$, and $b_l \ne c_l$.

If $k < l$, then $a_i = b_i = c_i$ for all $i = 1, 2, \ldots, k - 1$, $a_k\ R\ b_k$, $b_k = c_k$ (in which case $a_k\ R\ c_k$), and $a_k \ne c_k$ (since $a_k \ne b_k$). Thus if, $k < l$, then $s \preceq u$ by condition (2).

If $k = l$, then $b_k\ R\ c_k$ (in which case $a_k\ R\ c_k$ by transitivity of R) and $b_k \ne c_k$. It follows that $a_k \ne c_k$ *[for if $a_k = c_k$, then $a_k\ R\ b_k$ and $b_k\ R\ a_k$, which implies that $a_k = b_k$ (since R is a partial order) and contradicts the fact that $a_k \ne b_k$].* Thus if $k = l$, then $s \preceq u$ by condition (2).

If $k > l$, then $a_i = b_i = c_i$ for all $i = 1, 2, \ldots, l - 1$, $a_l\ R\ c_l$ (because $b_l\ R\ c_l$ and $a_l = b_l$), $a_l \ne c_l$ (because $b_l \ne c_l$ and $a_l = b_l$). Thus if $k > l$, then $s \preceq u$ by condition (2).

Hence, regardless of whether $k < l$, $k = l$, or $k > l$, we conclude that $s \preceq u$.

The above arguments show that in all possible situations, if $s \preceq t$ and $t \preceq u$ then $s \preceq u$ *[as was to be shown].* Hence $\preceq$ *is* transitive.

Conclusion: Since $\preceq$ *is* reflexive, antisymmetric, and transitive, $\preceq$ *is* a partial order relation.

15. Proof:

   Suppose R is a relation on a set A and R is reflexive, symmetric, transitive, and anti-symmetric. We will show that R is the identity relation on A.

   First note that for all x and y in A, if $x\ R\ y$ then, because R is symmetric, $y\ R\ x$. But then, because R is also anti-symmetric $x = y$. Thus for all x and y in A, if $x\ R\ y$ then $x = y$.

   This argument, however, does not prove that R is the identity relation on A because the conclusion would also follow from the hypothesis (by default) in the case where $A \ne \emptyset$ and $R = \emptyset$.

   But when $A \ne \emptyset$, it is impossible for R to equal $\emptyset$ because R is reflexive, which means that $x\ R\ x$ for every x in A.

   Thus every element in A is related by R to itself, and no element in A is related to anything other than itself. It follows that R is the identity relation on A.

27. greatest element: (1,1)   least element: (0,0)

   maximal elements: (1,1)   minimal elements: (0,0)

30. *c.* no greatest element and no least element

    d. greatest element: 9   least element: 1

33. *A* is not totally ordered by the given relation because $9 \nmid 12$ and $12 \nmid 9$.

36. $\{2, 4, 12, 24\}$ or $\{3, 6, 12, 24\}$

45. One such total order is 3, 9, 2, 6, 18, 4, 12, 8.

48. One such total order is $\emptyset, \{a\}, \{b\}, \{c\}, \{d\}, \{a,b\}, \{a,c\}, \{a,d\}, \{b,c\}, \{b,d\}, \{c,d\},$
    $\{a,b,c\}, \{a,b,d\}, \{a,c,d\}, \{b,c,d\}, \{a,b,c,d\}.$

51. *a.* 33 hours

Review Guide: Chapter 8

Definitions: How are the following terms defined?

- congruence modulo 2 relation *(p. 443)*
- inverse of a relation from a set A to a set B *(p. 444)*
- relation on a set *(p. 446)*
- directed graph of a relation on a set *(p. 446)*
- n-ary relation (and binary, ternary, quaternary relations) *(p. 447)*
- reflexive, symmetric, and transitive properties of a relation on a set *(p. 450)*
- congruence modulo 3 relation *(p. 455)*
- transitive closure of a relation on a set *(p. 457)*
- equivalence relation on a set *(p. 462)*
- equivalence class *(p. 465)*
- congruence modulo n relation *(p. 473)*
- representative of an equivalence class *(p. 472)*
- m is congruent to n modulo d *(p. 473)*
- plaintext and cyphertext *(p. 478)*
- residue of a modulo n *(p. 481)*
- complete set of residues modulo n *(p. 481)*
- d is a linear combination of a and b *(p. 486)*
- a and b are relatively prime; $a_1, a_2, \ldots, a_n$ are pairwise relatively prime *(p. 488)*
- an inverse of a modulo n *(p. 489)*
- antisymmetric relation *(p. 499)*
- partial order relation *(p. 500)*
- lexicographic order *(p. 502)*
- Hasse diagram *(p. 503)*
- a and b are comparable *(p. 505)*
- poset *(p. 506)*
- total order relation *(p. 506)*
- chain, length of a chain *(p. 506)*
- maximal element, greatest element, minimal element, least element *(p. 507)*
- topological sorting *(p. 507)*
- compatible partial order relations *(p. 508)*
- PERT and CPM *(p. 510)*
- critical path *(p. 512)*

Properties of Relations on Sets and Equivalence Relations

- How do you show that a relation on a finite set is reflexive? symmetric? transitive? *(pp. 450-452)*
- How do you show that a relation on an infinite set is reflexive? symmetric? transitive? *(pp. 453-456)*
- How do you show that a relation on a set is not reflexive? not symmetric? not transitive? *(pp. 451-454)*
- How do you find the transitive closure of a relation? *(p. 457)*
- What is the relation induced by a partition of a set? *(p. 460)*
- Given an equivalence relation on a set A, what is the relationship between the distinct equivalence classes of the relation and subsets of the set A? *(p. 469)*
- Given an equivalence relation on a set A and an element a in A, how do you find the equivalence class of a? *(pp. 465-467, 470-472)*
- In what way are rational numbers equivalence classes? *(pp. 473-474)*

Cryptography

- How does the Caesar cipher work? *(p. 478)*
- If a, b, and n are integers with $n > 1$, what are some different ways of expressing the fact that $n \mid (a - b)$? *(p. 480)*
- How do you reduce a number modulo n? *(p. 481)*
- If n is an integer with $n > 1$, is congruence modulo n an equivalence relation on the set of all integers? *(p. 481)*
- How do you add, subtract, and multiply integers modulo an integer $n > 1$? *(p. 482)*
- What is an efficient way to compute a^k where a is an integer with $a > 1$ and k is a large integer? *(pp. 484-485)*
- How do you express the greatest common divisor of two integers as a linear combination of the integers? *(p. 487)*
- When can you find an inverse modulo n for a positive integer a, and how do you find it? *(pp. 488-489)*
- How do you encrypt and decrypt messages using RSA cryptography? *(pp. 491-492)*
- What is Euclid's lemma? How is it proved? *(p. 492)*
- What is Fermat's little theorem? How is it proved? *(p. 494)*
- Why does the RSA cipher work? *(pp. 494-496)*

Partial Order Relations

- How do you show that a relation on a set is or is not antisymmetric? *(pp. 499-500)*
- If A is a set with a partial order relation R, S is a set of strings over A, and a and b are in S, how do you show that $a \preceq b$, where $\preceq$ denotes the lexicographic ordering of S? *(p. 502)*
- How do you construct the Hasse diagram for a partial order relation? *(p. 503)*
- How do you find a chain in a partially ordered set? *(p. 506)*
- Given a set with a partial order, how do you construct a topological sorting for the elements of the set? *(p. 508)*
- Given a job scheduling problem consisting of a number of tasks, some of which must be completed before others can be begun, how can you use a partial order relation to determine the minimum time needed to complete the job? *(pp. 511-512)*

Chapter 9: Counting and Probability

The primary aim of this chapter is to foster intuitive understanding for fundamental principles of counting and probability and an ability to apply them in a wide variety of situations. It is helpful to get into the habit of beginning a counting problem by listing (or at least imagining) some of the objects you are trying to count. If you see that all the objects to be counted can be matched up with the integers from m to n inclusive, then the total is $n - m + 1$ (Section 9.1). If you see that all the objects can be produced by a multi-step process, then the total can be found by counting the distinct paths from root to leaves in a possibility tree that shows the outcomes of each successive step (Section 9.2). And if each step of the process can be performed in a fixed number of ways (regardless of how the previous steps were performed), then the total can be calculated by applying the multiplication rule (Section 9.2).

If the objects to be counted can be separated into disjoint categories, then the total is just the sum of the subtotals for each category (Section 9.3). And if the categories are not disjoint, the total can be counted using the inclusion/exclusion rule (Section 9.3). If the objects to be counted can be represented as all the subsets of size r of a set with n elements, then the total is $\binom{n}{r}$ for which there is a computational formula (Section 9.5). Finally if the objects can be represented as all the multisets of size r of a set with n elements, then the total is $\binom{n+r-1}{r}$ (Section 9.6).

Section 9.4 introduces the pigeonhole principle, which provides a way to answer questions about how many of a certain object are needed to guarantee certain results and is used to show that certain results are guaranteed if a certain number of objects are present. The section includes the reasoning for why every rational number has a decimal expansion that either terminates or repeats.

Pascal's formula and the binomial theorem are discussed in Section 9.7. Each is proved both algebraically and combinatorially. Pascal's formula and the binomial theorem are discussed in Section 9.7. Each is proved both algebraically and combinatorially. Exercise 28 of Section 9.7 is intended to help you see how Pascal's formula is applied in the algebraic proof of the binomial theorem.

Sections 9.8 and 9.9 develop the axiomatic theory of probability through the concepts of expected value, conditional probability, independence, and Bayes' theorem. Exercise 20 of Section 9.1 can be solved directly by reasoning about the sample space, but it can also be solved using conditional probability, which is discussed in Section 9.9.

Section 9.1

6. $\{2\clubsuit, 3\clubsuit, 4\clubsuit, 2\diamondsuit, 3\diamondsuit, 4\diamondsuit, 2\heartsuit, 3\heartsuit, 4\heartsuit, 2\spadesuit, 3\spadesuit, 4\spadesuit\}$ Probability $= 12/52 = 3/13 \cong 23.1\%$

12. *b.* (ii) $\{GGB, GBG, BGG, GGG\}$ Probability $= 4/8 = 1/2 = 50\%$

    (iii) $\{BBB\}$ Probability $= 1/8 = 12.5\%$

15. The methods used to compute the probabilities in exercises 12, 13, and 14 are exactly the same as those in exercise 11. The only difference in the solutions are the symbols used to denote the outcomes; the probabilities are identical. These exercises illustrate the fact that computing various probabilities that arise in connection with tossing a coin is mathematically identical to computing probabilities in other, more realistic situations. So if the coin tossing model is completely understood, many other probabilities can be computed without difficulty.

27. Let k be the 62nd element in the array. By Theorem 9.1.1, $k - 29 + 1 = 62$, so $k = 62 + 29 - 1 = 90$. Thus the 62nd element in the array is $B[90]$.

30. 1   2   3   4   5   6   ...   998   999   1000   1001
        $\updownarrow$ $\updownarrow$ $\updownarrow$ $\updownarrow$ $\updownarrow$
        $2 \cdot 1$ $2 \cdot 2$ $2 \cdot 3$ $2 \cdot 499$ $2 \cdot 500$

    The diagram above shows that there are as many even integers between 1 and 1001 as there are integers from 1 to 500 inclusive. There are 500 such integers.

125

33. <u>Proof (by mathematical induction)</u>: Let the property $P(n)$ be the sentence

    The number of integers from m to n inclusive is $n - m + 1$. $\leftarrow P(n)$

We will prove by mathematical induction that the property is true for all integers $n \geq m$.

Show that $P(m)$ is true: $P(m)$ is true because there is just one integer, namely m, from m to m inclusive. Substituting m in place of n in the formula $n - m + 1$ gives $m - m + 1 = 1$, which is correct.

Show that for all integers $k \geq m$, if $P(k)$ is true then $P(k + 1)$ is true: Let k be any integer with $k \geq m$ and suppose that

    The number of integers from m to k inclusive is $k - m + 1$. $\leftarrow$ $P(k)$
    inductive hypothesis

We must show that

    The number of integers from m to $k + 1$ inclusive is $(k + 1) - m + 1$. $\leftarrow P(k + 1)$

Consider the sequence of integers from m to $k + 1$ inclusive:

$$\underbrace{m, \quad m + 1, \quad m + 2, \quad \ldots, \quad k,}_{k - m + 1} \quad (k + 1).$$

By inductive hypothesis there are $k - m + 1$ integers from m to k inclusive. So there are $(k - m + 1) + 1$ integers from m to $k + 1$ inclusive. But $(k - m + 1) + 1 = (k + 1) - m + 1$. So there are $(k + 1) - m + 1$ integers from m to $k + 1$ inclusive *[as was to be shown]*.

Section 9.2

12. *b.* Think of creating a string of hexadecimal digits that satisfies the given requirements as a 6-step process.

    Step 1: Choose the first hexadecimal digit. It can be any hexadecimal digit from 4 through D (which equals 13). There are $13 - 4 + 1 = 10$ of these.

    Steps 2–5: Choose the second through the fifth hexadecimal digits. Each can be any one of the 16 hexadecimal digits.

    Step 6: Choose the last hexadecimal digit. It can be any hexadecimal digit from 2 through E (which equals 14). There are $14 - 2 + 1 = 13$ of these.

    So the total number of the specified hexadecimal numbers is $10 \cdot 16 \cdot 16 \cdot 16 \cdot 16 \cdot 13 = 8,519,680$.

15. Think of creating combinations that satisfy the given requirements as multi-step processes in which each of steps 1-3 is to choose a number from 1 to 30, inclusive.

    a. Because there are 30 choices of numbers in each of steps 1–3, there are $30^3 = 27,000$ possible combinations for the lock.

    b. In this case we are given that no number may be repeated. So there are 30 choices for step 1, 29 for step 2, and 28 for step 3. Thus there are $30 \cdot 29 \cdot 28 = 24,360$ possible combinations for the lock.

18. *b.* Constructing a PIN that is obtainable by the same keystroke sequence as 5031 can be thought of as the following four-step process:

Step 1: Choose either the digit 5 or one of the three letters on the same key as the digit 5.

Step 2: Choose the digit 0.

Step 3: Choose the digit 3 or one of the three letters on the same key as the digit 3.

Step 4: Choose either the digit 1 or one of the two letters on the same key as the digit 1.

There are four ways to perform steps 1 and 3, one way to perform step 2, and three ways to perform step 4. So by the multiplication rule there are $4 \cdot 1 \cdot 4 \cdot 3 = 48$ different PINs that are keyed the same as 5031.

c. Constructing a numeric PIN with no repeated digit can be thought of as the following four-step process. Steps 1–4 are to choose the digits in position 1–4 (counting from the left). Because no digit may be repeated, there are 10 ways to perform step one, 9 ways to perform step two, 8 ways to perform step three, and 7 ways to perform step four. Thus the number of numeric PINs with no repeated digit is $10 \cdot 9 \cdot 8 \cdot 7 = 5040$.

21. *a.* There are 2^{mn} relations from A to B because a relation from A to B is any subset of $A \times B$, $A \times B$ is a set with mn elements (since A has m elements and B has n elements), and the number of subsets of a set with mn elements is 2^{mn} (by Theorem 6.3.1).

b. In order to define a function from A to B we must specify exactly one image in B for each of the m elements in A. So we can think of constructing a function from A to B as an m-step process, where step i is to choose an image for the ith element of A (for $i = 1, 2, \ldots, m$). Because there are n choices of image for each of the m elements, by the multiplication rule, the total number of functions is $\underbrace{n \cdot n \cdot n \cdots n}_{m \text{ factors}} = n^m$.

c. The fraction of relations from A to B that are functions is $\dfrac{n^m}{2^{nm}} = \left(\dfrac{n}{2^n}\right)^m$

30. *a.* Call one of the integers r and the other s. Since r and s have no common factors, if p_i is a factor of r, then p_i is not a factor of s.

So for each $i = 1, 2, \ldots, m$, either $p_i{}^{k_i}$ is a factor of r or $p_i{}^{k_i}$ is a factor of s.

Thus, constructing r can be thought of as an m-step process in which step i is to decide whether $p_i{}^{k_i}$ is a factor of r or not.

There are two ways to perform each step, and so the number of different possible r's is 2^m.

Observe that once r is specified, s is completely determined because $s = n/r$.

Hence the number of ways n can be written as a product of two positive integers rs which have no common factors is 2^m. Note that this analysis assumes that order matters because, for instance, $r = 1$ and $s = n$ will be counted separately from $r = n$ and $s = 1$.

b. Each time that we can write n as rs, where r and s have no common factors, we can also write $n = sr$. So if order matters, there are twice as many ways to write n as a product of two integers with no common factors as there are if order does not matter. Thus if order does not matter, there are $2^m/2 = 2^{m-1}$ ways to write n as a product of two integers with no common factors.

33. *a.* The number of ways the 6 people can be seated equals the number or permutations of a set of 6 elements, namely, $6! = 720$.

b. Assuming that the row is bounded by two aisles, arranging the people in the row can be regarded as the following 2-step process:

Step 1: Choose the aisle seat for the doctor. *[There are 2 ways to do this.]*

Step 2: Choose an ordering for the remaining people. *[There are 5! ways to do this.]*

Thus, by the multiplication rule, the answer is $2 \cdot 5! = 240$.

(If it is assumed that one end of the row is against a wall, then there is only one aisle seat and the answer is $5! = 120$.)

c. Each married couple can be regarded as a single item, so the number of ways to order the 3 couples is $3! = 6$.

36. *stu, stv, sut, suv, svt, svu, tsu, tsv, tus, tuv, tvs, tvu, ust, usv, uts, utv, uvs, uvt, vst, vsu, vts, vtu, vus, vut*

39. *b.* $P(9,6) = 9!/(9-6)! = 9!/3! = 9 \cdot 8 \cdot 7 \cdot 6 \cdot 5 \cdot 4 = 60{,}480$

    d. $P(7,4) = 7!/(7-4)! = 7!/3! = 7 \cdot 6 \cdot 5 \cdot 4 = 840$

42. <u>Proof 1</u>: Let n be any integer such that $n \geq 3$. By the first version of the formula in Theorem 9.2.3,

$$
\begin{aligned}
P(n+1,3) - P(n,3) &= (n+1)n(n-1) - n(n-1)(n-2) \\[2mm]
&= n(n-1)[(n+1) - (n-2)] \\[2mm]
&= n(n-1)(n+1-n+2) \\[2mm]
&= 3n(n-1) \\[2mm]
&= 3P(n,2).
\end{aligned}
$$

<u>Proof 2</u>: Let n be any integer such that $n \geq 3$. By the second version of the formula in Theorem 9.2.3,

$$
\begin{aligned}
P(n+1,3) - P(n,3) &= \frac{(n+1)!}{((n+1)-3)!} - \frac{n!}{(n-3)!} \\[3mm]
&= \frac{(n+1)!}{(n-2)!} - \frac{n!}{(n-3)!} \\[3mm]
&= \frac{(n+1) \cdot n!}{(n-2)!} - \frac{(n-2) \cdot n!}{(n-2) \cdot (n-3)!} \\[3mm]
&= \frac{n!((n+1) - (n-2))}{(n-2)!} \\[3mm]
&= \frac{n!}{(n-2)!} \cdot 3 \\[3mm]
&= 3P(n,2).
\end{aligned}
$$

45. <u>Proof (by mathematical induction)</u>: Let the property $P(n)$ be the sentence

The number of permutations of a set with n elements is $n!$. $\leftarrow P(n)$

We will prove by mathematical induction that the property is true for all integers $n \geq 1$.

Show that $P(1)$ is true: $P(1)$ is true because if a set consists of one element there is just one way to order it, and $1! = 1$.

Show that for all integers $k \geq 1$, if $P(k)$ is true then $P(k+1)$ is true: Let k be any integer with $k \geq 1$ and suppose that

The number of permutations of a set with k elements is $k!$. $\leftarrow$ $\begin{array}{l}P(k)\\ \text{inductive hypothesis}\end{array}$

We must show that

number of permutations of a set with k elements is $(k+1)!$. $\leftarrow P(k+1)$

Let X be a set with $k + 1$ elements. The process of forming a permutation of the elements of X can be considered a two-step operation as follows:

Step 1: Choose the element to write first.

Step 2: Write the remaining elements of X in some order.

Since X has $k+1$ elements, there are $k+1$ ways to perform step 1, and by inductive hypothesis there are $k!$ ways to perform step 2. Hence by the multiplication rule there are $(k + 1)k! = (k + 1)!$ ways to form a permutation of the elements of X. But this means that there are $(k + 1)!$ permutations of X *[as was to be shown]*.

Section 9.3

6. *a*. For simplicity, start by assuming that a blank plate is allowed. Then the number of ways to construct a license plate can be thought of as a 2-step process, where step 1 is to choose the letters for the initial portion of the plate and Step 2 is to choose the digits for the second portion of the plate. Because anywhere from 0 to 3 letters may be chosen for the initial portion of the plate, by the addition rule, the number of ways to choose the letters for the initial portion of the plate is

$$\begin{pmatrix} \text{the number} \\ \text{of choices of} \\ \text{of 0 letters} \end{pmatrix} + \begin{pmatrix} \text{the number} \\ \text{of choices of} \\ \text{of 1 letter} \end{pmatrix} + \begin{pmatrix} \text{the number} \\ \text{of choices of} \\ \text{of 2 letters} \end{pmatrix} + \begin{pmatrix} \text{the number} \\ \text{of choices of} \\ \text{of 3 letters} \end{pmatrix}.$$

Because there are 26 letters in the alphabet, there is only one way to choose 0 letters, and 26 ways to choose one letter. The number of ways to choose two or three letters is computed using the multiplication rule. For example, choosing three letters can be thought of as a 3-step process: step 1 is to fill in the first letter, step 2 is to fill in the second letter, and step 3 is to fill in the third letter. Thus the number of ways to choose three letters is 26^3. Similarly, the number of ways to choose two letters is 26^2. It follows that the number of ways to choose from 0 to 3 letters is

$$1 + 26 + 26^2 + 26^3.$$

The same kind of reasoning can be applied to compute the number of ways to choose the digits for the second portion of the license plate, namely

$$\begin{pmatrix} \text{the number} \\ \text{of choices of} \\ \text{of 0 digits} \end{pmatrix} + \begin{pmatrix} \text{the number} \\ \text{of choices of} \\ \text{of 1 digit} \end{pmatrix} + \begin{pmatrix} \text{the number} \\ \text{of choices of} \\ \text{of 2 digits} \end{pmatrix} + \begin{pmatrix} \text{the number} \\ \text{of choices of} \\ \text{of 3 digits} \end{pmatrix} + \begin{pmatrix} \text{the number} \\ \text{of choices of} \\ \text{of 4 digits} \end{pmatrix}.$$

Because there are ten digits, there is only one way to choose 0 digits, and 10 ways to choose one digit. The number of ways to choose two, three, or four digits is computed using the multiplication rule. For example, choosing three digits can be thought of as a 3-step process: step 1 is to fill in the first digit, step 2 is to fill in the second digit, and step 3 is to fill in the third digit. Thus the number of ways to choose three digits is 26^3. Similarly, the number of ways to choose two digits is 26^2, and the number of ways to choose four digits is 26^4. It follows that the number of ways to choose from 0 to 4 digits is

$$1 + 10 + 10^2 + 10^3 + 10^4.$$

Since each choice of from 0 to 3 letters can be paired with each choice of from 0 to 4 digits, by the multiplication rule, the number of ways to choose from 0 to 3 letters to place in the initial portion of the license plate and from 0 to 4 digits to place in the final portion of the license plate is the product

$$(1 + 26 + 26^2 + 26^3)(1 + 10 + 10^2 + 10^3 + 10^4) = 203,097,969.$$

However, this number includes the blank plate, which is not allowed. So, by the difference rule, the total number of license plates is $203,097,969 - 1 = 203,097,968$.

9. *b.* On the ith iteration of the outer loop, there are i iterations of the inner loop, and this is true for each $i = 1, 2, \ldots, n$. Therefore, the total number of iterations of the inner loop is $1 + 2 + 3 + \cdots + n = n(n+1)/2$.

12. *a.* The number of ways to arrange the 6 letters of the word *THEORY* in a row is $6! = 720$

b. When the *TH* in the word *THEORY* are treated as an ordered unit, there are only 5 items to arrange, *TH, E, O, R,* and *Y.* and so there are 5! orderings. Similarly, there are 5! orderings for the symbols *HT, E, O, R,* and *Y.* Thus, by the addition rule, the total number of orderings is $5! + 5! = 120 + 120 = 240$.

15. The set of all possible identifiers may be divided into 30 non-overlapping subsets depending on the number of characters in the identifier. Constructing one of the identifiers in the kth subset can be regarded as a k-step process, where each step consists in choosing a symbol for one of the characters (say, going from left to right). Because the first character must be a letter, there are 26 choices for step 1, and because subsequent letters can be letters or digits or underscores there are 37 choices for each subsequent step. By the addition rule, we add up the number of identifiers in each subset to obtain a total. But because 82 of the resulting strings cannot be used as identifiers, by the difference rule, we subtract 82 from the total to obtain the final answer. Thus we have

$$(26 + 26 \cdot 37 + 26 \cdot 37^2 + \cdots + 26 \cdot 37^{29}) - 82 = 26(1 + 37 + 37^2 + \cdots + 37^{29}) - 82$$

$$= 26 \cdot \sum_{k=0}^{29} 37^k - 82 = 26\left(\frac{37^{30} - 1}{37 - 1}\right) - 82 \cong 8.030 \times 10^{46} \cong.$$

18. *b.* <u>Proof:</u> Let A and B be events in a sample space S. By the inclusion/exclusion rule (Theorem 9.3.3), $N(A \cup B) = N(A) + N(B) - N(A \cap B)$. So by the equally likely probability formula,

$$P(A \cup B) = \frac{N(A \cup B)}{N(S)} = \frac{N(A) + N(B) - N(A \cap B)}{N(S)} = \frac{N(A)}{N(S)} + \frac{N(B)}{N(S)} - \frac{N(A \cap B)}{N(S)}$$

$$= P(A) + P(B) - P(A \cap B).$$

21. Call the employees $U, V, W, X, Y,$ and Z, and suppose that U and V are the married couple. Let A be the event that U and V have adjacent desks. Since the desks of U and V can be adjacent either in the order UV or in the order VU, the number of desk assignments with U and V adjacent is the same as the sum of the number of permutations of the symbols $\boxed{UV}$, W, X, Y, Z plus the number of permutations of the symbols $\boxed{VU}$, W, X, Y, Z. By the multiplication rule each of these is 5!, and so by the addition rule the sum is $2 \cdot 5!$. Since the total number of permutations of U, V, W, X, Y, Z is 6!,

$$P(A) = 2 \cdot \frac{5!}{6!} = \frac{2}{6} = \frac{1}{3}.$$

Hence by the formula for the probability of the complement of an event,

$$P(A^c) = 1 - P(A) = 1 - \frac{1}{3} = \frac{2}{3}.$$

So the probability that the married couple have nonadjacent desks is 2/3.

24. *a.* Let A and B be the sets of all integers from 1 through 1,000 that are multiples of 2 and 9 respectively. Then $N(A) = 500$ and $N(B) = 111$ (because $9 = 9 \cdot 1$ is the smallest integer in B and $999 = 9 \cdot 111$ is the largest). Also $A \cap B$ is the set of all integers from 1 through 1,000 that are multiples of 18, and $N(A \cap B) = 55$ (because $18 = 18 \cdot 1$ is the smallest integer in $A \cap B$ and $990 = 18 \cdot 55$ is the largest). It follows from the inclusion/exclusion rule that the number of integers from 1 through 1,000 that are multiples of 2 or 9 equals

$$N(A \cup B) = N(A) + N(B) - N(A \cup B) = 500 + 111 - 55 = 556.$$

b. The probability is $556/1000 = 55.6\%$.

c. $1000 - 556 = 444$

27. *a.* Let k be an integer with $k \geq 3$. The set of bit strings of length k that do not contain the pattern 101 can be partitioned into $k + 1$ subsets: the subset of strings that start with 0 and continue with any bit string of length $k - 1$ not containing 101 *[there are a_{k-1} of these]*, the subset of strings that start with 100 and continue with any bit string of length $k - 3$ not containing 101 *[there are a_{k-3} of these]*, the subset of strings that start with 1100 and continue with any bit string of length $k - 4$ not containing 101 *[there are a_{k-4} of these]*, the subset of strings that start with 11100 and continue with any bit string of length $k - 5$ not containing 101 *[there are a_{k-5} of these]*, until the following subset of strings is obtained: $\{\underbrace{11\ldots1}_{k-3\ 1's}001, \underbrace{11\ldots1}_{k-3\ 1's}000\}$ *[there are 2 of these and a_1 equals 2]*. In addition, the three single-element sets $\{\underbrace{11\ldots1}_{k-2\ 1's}00\}$, $\{\underbrace{11\ldots1}_{k-1\ 1's}0\}$, and $\{\underbrace{11\ldots1}_{k-1\ 1's}1\}$ are in the partition, and since $a_0 = 1$ (because the only bit string of length zero that satisfies the condition is ϵ), $3 = a_0 + 2$. Thus by the addition rule,

$$a_k = a_{k-1} + a_{k-3} + a_{k-4} + \cdots + a_1 + a_0 + 2.$$

b. By part (a), if $k \geq 4$,

$$
\begin{aligned}
a_k &= a_{k-1} + a_{k-3} + a_{k-4} + \cdots + a_1 + a_0 + 2 \\
a_{k-1} &= a_{k-2} + a_{k-4} + a_{k-5} + \cdots + a_1 + a_0 + 2.
\end{aligned}
$$

Subtracting the second equation from the first gives

$$
\begin{aligned}
a_k - a_{k-1} &= a_{k-1} + a_{k-3} - a_{k-2} \\
\Rightarrow \qquad a_k &= 2a_{k-1} + a_{k-3} - a_{k-2}. \text{ (Call this equation (*).)}
\end{aligned}
$$

Note that $a_2 = 4$ (because all four bit strings of length 2 satisfy the condition) and $a_3 = 7$ (because all eight bit strings of length 3 satisfy the condition except 101). Thus equation (*) is also satisfied when $k = 3$ because in that case the right-hand side of the equation becomes $2a_2 + a_0 - a_1 = 2 \cdot 4 + 1 - 2 = 7$, which equals the left-hand side of the equation.

30. To get a sense of the problem, we compute s_4 directly. If there are four seats in the row, there can be a single student in any one of the four seats or there can be a pair of students in seats 1&3, 1&4, or 2&4. No other arrangements are possible because with more than two students, two would have to sit next to each other. Thus $s_4 = 4 + 3 = 7$. In general, if there are k chairs in a row, then

$$
\begin{aligned}
s_k \ = \ & s_{k-1} && \text{(the number of ways a nonempty set of students can sit} \\
& && \text{in the row with no two students adjacent and chair } k \text{ empty)} \\[4pt]
& +s_{k-2} && \text{(the number of ways students can sit in the row with chair } k \\
& && \text{occupied, chair } k-1 \text{ empty, and chairs 1 through} \\
& && k-2 \text{ occupied by a nonempty set of students in such a} \\
& && \text{way that no two students are adjacent)} \\[4pt]
& +1 && \text{(for the seating in which chair } k \text{ is occupied} \\
& && \text{and all the other chairs are empty} \\[4pt]
= \ & s_{k-1} + s_{k-2} + 1 && \text{for all integers } k \geq 3.
\end{aligned}
$$

33. *c.*

e. 1   *f.* 17

36. *a.* by the double complement law and the difference rule   *b.* by De Morgan's law

    c. by the inclusion/exclusion rule

39. Imagine each integer from 1 through 999,999 as a string of six digits with leading 0's included. For each $i = 1, 2, 3$, let A_i be the set of all integers from 1 through 999,999 that do not contain the digit i. We want to compute $N(A_1{}^c \cap A_2{}^c \cap A_3{}^c)$. By De Morgan's law,

$$A_1{}^c \cap A_2{}^c \cap A_3{}^c = (A_1 \cup A_2)^c \cap A_3{}^c = (A_1 \cup A_2 \cup A_3)^c = U - (A_1 \cup A_2 \cup A_3),$$

and so, by the difference rule,

$$N(A_1{}^c \cap A_2{}^c \cap A_3{}^c) = N(U) - N(A_1 \cup A_2 \cup A_3).$$

By the inclusion/exclusion rule,

$$N(A_1 \cup A_2 \cup A_3) = N(A_1) + N(A_2) + N(A_3) - N(A_1 \cap A_2) - N(A_1 \cap A_3) - N(A_2 \cap A_3) + N(A_1 \cap A_2 \cap A_3).$$

Now $N(A_1) = N(A_2) = N(A_3) = 9^6$ because in each case any of nine digits may be chosen for each character in the string (for A_i these are all the ten digits except i). Also each $N(A_i \cap A_j) = 8^6$ because in each case any of eight digits may be chosen for each character of the string (for $A_i \cap A_j$ these are all the ten digits except i and j). Similarly, $N(A_1 \cap A_2 \cap A_3) = 7^6$ because any digit except 1, 2, and 3 may be chosen for each character in the string. Thus

$$N(A_1 \cup A_2 \cup A_3) = 3 \cdot 9^6 - 3 \cdot 8^6 + 7^6,$$

and so, by the difference rule,

$$N(A_1{}^c \cap A_2{}^c \cap A_3{}^c) = N(U) - N(A_1 \cup A_2 \cup A_3) = 10^6 - (3 \cdot 9^6 - 3 \cdot 8^6 + 7^6) = 74,460.$$

42. *a.* $g_3 = 1$, $g_4 = 1$, $g_5 = 2$ (*LWLLL* and *WWLLL*)

    b. $g_6 = 4$ (*WWWLLL*, *WLWLLL*, *LWWLLL*, *LLWLLL*)

    c. If $k \geq 6$, then any sequence of k games must begin with exactly one of the possibilities: W, LW, or LLW. The number of sequences of k games that begin with W is g_{k-1} because the succeeding $k - 1$ games can consist of any sequence of wins and losses except that the first sequence of three consecutive losses occurs at the end. Similarly, the number of sequences of k games that begin with LW is g_{k-2} and the number of sequences of k games that begin with LLW is g_{k-3}. Therefore, $g_k = g_{k-1} + g_{k-2} + g_{k-3}$ for all integers $k \geq 6$.

45. <u>Proof (by mathematical induction):</u> Let the property $P(k)$ be the sentence

> If a finite set A equals the union of k distinct mutually
> disjoint subsets subsets $A_1, A_2, \ldots, A_k$, then $\leftarrow P(k)$
> $N(A) = N(A_1) + N(A_2) + \cdots + N(A_k).$

We will prove by mathematical induction that $P(k)$ is true for all integers $k \geq 1$.

Show that $P(1)$ is true: $P(1)$ is true because if a finite set A equals the "union" of one subset A_1, then $A = A_1$, and so $N(A) = N(A_1)$.

Show that for all integers $i \geq 1$, if $P(i)$ is true then $P(i+1)$ is true: Let i be any integer with $i \geq 1$ and suppose that

If a finite set A equals the union of i distinct mutually disjoint subsets subsets $A_1, A_2, \ldots, A_i$, then $N(A) = N(A_1) + N(A_2) + \cdots + N(A_i)$. $\leftarrow$ $P(i)$ inductive hypothesis

We must show that

If a finite set A equals the union of $i+1$ distinct mutually disjoint subsets subsets $A_1, A_2, \ldots, A_{i+1}$, then $N(A) = N(A_1) + N(A_2) + \cdots + N(A_{i+1})$. $\leftarrow$ $P(i+1)$

Let A be a finite set that equals the union of $i+1$ distinct mutually disjoint subsets $A_1, A_2, \ldots, A_{i+1}$. Then $A = A_1 \cup A_2 \cup \cdots \cup A_{i+1}$ and $A_i \cap A_j = \emptyset$ for all integers i and j with $i \neq j$.

Let B be the set $A_1 \cup A_2 \cup \cdots \cup A_i$. Then $A = B \cup A_{i+1}$ and $B \cap A_{i+1} = \emptyset$.

[For if $x \in B \cap A_{i+1}$, then $x \in A_1 \cup A_2 \cup \cdots \cup A_i$ and $x \in A_{i+1}$, which implies that $x \in A_j$, for some j with $1 \leq j \leq i$, and $x \in A_{i+1}$. But A_j and A_i are disjoint. Thus no such x exists.]

Hence A is the union of the two mutually disjoint sets B and A_{i+1}. Since B and A_{i+1} have no elements in common, the total number of elements in $B \cup A_{i+1}$ can be obtained by first counting the elements in B, next counting the elements in A_{i+1}, and then adding the two numbers together.

It follows that $N(B \cup A_{i+1}) = N(B) + N(A_{i+1})$ which equals $N(A_1) + N(A_2) + \cdots + N(A_i) + N(A_{i+1})$ by inductive hypothesis. Hence $P(i+1)$ is true *[as was to be shown]*.

48. Proof (by mathematical induction): Let the property $P(n)$ be the general inclusion/exclusion rule. We will prove by mathematical induction that $P(n)$ is true for all integers $n \geq 2$.
Show that $P(2)$ is true: $P(2)$ was proved in one way in the text preceding Theorem 9.3.3 and in another way in the solution to exercise 46.
Show that for all integers $r \geq 2$, if $P(r)$ is true then $P(r+1)$ is true: Let r be any integer with $r \geq 2$ and suppose that the general inclusion/exclusion rule holds for any collection of r finite sets. (This is the inductive hypothesis.) Let $A_1, A_2, \ldots, A_{r+1}$ be finite sets. Then

$N(A_1 \cup A_2 \cup \cdots \cup A_{r+1})$

$= N(A_1 \cup (A_2 \cup A_3 \cup \cdots \cup A_{r+1}))$ by the associative law for $\cup$

$= N(A_1) + N(A_2 \cup A_3 \cup \cdots \cup A_{r+1}) - N(A_1 \cap (A_2 \cup A_3 \cup \cdots \cup A_{r+1}))$ by the inclusion/exclusion rule for two sets

$= N(A_1) + N(A_2 \cup A_3 \cup \cdots \cup A_{r+1}) - N((A_1 \cap A_2) \cup (A_1 \cap A_3) \cup \cdots \cup (A_1 \cap A_{r+1}))$ by the generalized distributive law for sets (exercise 37, Section 6.2)

$= N(A_1) + \left(\sum_{2 \leq i \leq r+1} N(A_i) - \sum_{2 \leq i < j \leq r+1} N(A_i \cap A_j) \right.$
$\left. + \sum_{2 \leq i < j < k \leq r+1} N(A_i \cap A_j \cap A_k) - \cdots + (-1)^{r+1} N(A_2 \cap A_3 \cap \cdots \cap A_{r+1}) \right)$
$- \left(\sum_{2 \leq i \leq r+1} N(A_1 \cap A_i) - \sum_{2 \leq i < j \leq r+1} N((A_1 \cap A_i) \cap (A_1 \cap A_j)) + \cdots \right.$
$\left. + (-1)^{r+1} N((A_1 \cap A_2) \cap (A_1 \cap A_3) \cap \cdots \cap (A_1 \cap A_{r+1})) \right)$ by inductive hypothesis

$$= N(A_1) + \left(\sum_{2 \leq i \leq r+1} N(A_i) - \sum_{2 \leq i < j \leq r+1} N(A_i \cap A_j) \right.$$

$$\left. + \sum_{2 \leq i < j < k \leq r+1} N(A_i \cap A_j \cap A_k) - \cdots + (-1)^{r+1} N(A_2 \cap A_3 \cap \cdots \cap A_{r+1}) \right)$$

$$- \left(\sum_{2 \leq i \leq r+1} N(A_1 \cap A_i) - \sum_{2 \leq i < j \leq r+1} N(A_1 \cap A_i \cap A_j) + \cdots \right.$$

$$\left. + (-1)^{r+1} N(A_1 \cap A_2 \cap A_3 \cap \cdots \cap A_{r+1}) \right)$$

$$= \sum_{1 \leq i \leq r+1} N(A_i) - \sum_{1 \leq i < j \leq r+1} N(A_i \cap A_j) + \sum_{1 \leq i < j < k \leq r+1} N(A_i \cap A_j \cap A_k)$$

$$- \cdots + (-1)^{r+2} N(A_1 \cap A_3 \cap \cdots \cap A_{r+1}).$$

[This is what was to be proved.]

Section 9.4

6. *a.* Yes.

   Solution 1: There are 6 possible remainders that can be obtained when an integer is divided by 7, namely 0, 1, 2, 3, 4, 5. Apply the pigeonhole principle, thinking of the 7 integers as the pigeons and the possible remainders as the pigeonholes. Each pigeon flies into the pigeonhole that is the remainder obtained when it is divided by 6. Since $7 > 6$, the pigeonhole principle says that at least two pigeons must fly into the same pigeonhole. So at least two of the numbers must have the same remainder when divided by 6.

   Solution 2: Let X be the set of seven integers and Y the set of all possible remainders obtained through division by 6, and consider the function R from X (the pigeons) to Y (the pigeonholes) defined by the rule: $R(n) = n \ mod \ 6$ (= the remainder obtained by the integer division of n by 6). Now X has 7 elements and Y has 6 elements (0, 1, 2, 3, 4, and 5). Hence by the pigeonhole principle, R is not one-to-one: $R(n_1) = R(n_2)$ for some integers n_1 and n_2 with $n_1 \neq n_2$. But this means that n_1 and n_2 have the same remainder when divided by 6.

   b. No. Consider the set $\{1, 2, 3, 4, 5, 6, 7\}$. This set has seven elements no two of which have the same remainder when divided by 8.

15. There are $n + 1$ even integers from 0 to $2n$ inclusive:

$$0 \,(= 2 \cdot \underline{0}), \ 2 \,(= 2 \cdot \underline{1}), \ 4 \,(= 2 \cdot \underline{2}), \ldots, \ 2n \,(= 2 \cdot \underline{n}).$$

So a maximum of $n + 1$ even integers can be chosen. Thus if at least $n + 2$ integers are chosen, one is sure to be odd. Similarly, there are n odd integers from 0 to $2n$ inclusive, namely

$$1 \,(= 2 \cdot \underline{1} - 1), \ 3 \,(= 2 \cdot \underline{2} - 1), \ldots, \ 2n - 1 \,(= 2 \cdot \underline{n} - 1).$$

It follows that if at least $n + 1$ integers are chosen, one is sure to be even.

(An alternative way to reach the second conclusion is to note that there are $2n + 1$ integers from 0 to $2n$ inclusive. Because $n + 1$ of them are even, the number of odd integers is $(2n + 1) - (n + 1) = n$.)

18. There are 15 distinct remainders that can be obtained through integer division by 15 (0, 1, 2, ..., 14). Hence at least 16 integers must be chosen in order to be sure that at least two have the same remainder when divided by 15.

21. The length of the repeating section of the decimal representation of 683/1493 is less than or equal to 1,492. The reason is that there are 1,492 nonzero remainders that can be obtained when a number is divided by 1,493. Thus, in the long-division process of dividing 683.0000... by 1,493, either some remainder is 0 and the decimal expansion terminates (in which case the

length of the repeating section is 0) or, only nonzero remainders are obtained and at some point within the first 1,492 successive divisions, a nonzero remainder is repeated. At that point the digits in the developing decimal expansion begin to repeat because the sequence of successive remainders repeats those previously obtained.

27. Yes. Let X be the set of 2,000 people (the pigeons) and Y the set of all 366 possible birthdays (the pigeonholes). Define a function $B \colon X \to Y$ by specifying that $B(x) = x$'s birthday. Now $2000 > 4 \cdot 366 = 1464$, and so by the generalized pigeonhole principle, there must be some birthday y such that $B^{-1}(y)$ has at least $4 + 1 = 5$ elements. Hence at least 5 people must share the same birthday.

30. Consider the maximum number of pennies that can be chosen without getting at least five from the same year. This maximum, which is 12, is obtained when four pennies are chosen from each of the three years. Hence at least thirteen pennies must be chosen to be sure of getting at least five from the same year.

33. <u>Proof:</u> Suppose A is a set of six positive integers each of which is less than 15. By Theorem 6.3.1, $\mathscr{P}(A)$, the power set of A, has $2^6 = 64$ elements, and so A has 63 nonempty subsets. Let k be the smallest number in the set A.

Given any nonempty subset of A, the sum of all the elements in the subset lies in the range from k through $k + 10 + 11 + 12 + 13 + 14 = k + 60$, and, by Theorem 9.1.1, there are $(k + 60) - k + 1 = 61$ integers in this range. Let S be the set of all possible sums of the elements that are in a nonempty subset of A. Then S has at most 61 elements.

Define a function F from the set of nonempty subsets of A to S as follows: For each nonempty subset X in A, let $F(X)$ be the sum of the elements of X. Because A has 63 nonempty subsets and S has 61 elements, the pigeonhole principle guarantees that F is not one-to-one. Thus there exist distinct nonempty subsets A_1 and A_2 of A such that $F(A_1) = F(A_2)$, which implies that the elements of A_1 add up to the same sum as the elements of A_2.

Note: In fact, it can be shown that it is always possible to find disjoint subsets of A with the same sum. To see why this is true, consider again the sets A_1 and A_2 found in the preceding proof. Then $A_1 \neq A_2$ and $F(A_1) = F(A_2)$. By definition of F, $F(A_1 - A_2) + F(A_1 \cap A_2) =$ the sum of the elements in $A_1 - A_2$ plus the sum of the elements in $A_1 \cap A_2$. But $A_1 - A_2$ and $A_1 \cap A_2$ are disjoint and their union is A_1. So $F(A_1 - A_2) + F(A_1 \cap A_2) = F(A_1)$. By the same reasoning, $F(A_2 - A_1) + F(A_1 \cap A_2) = F(A_2)$. Since $F(A_1) = F(A_2)$, we have that $F(A_1 - A_2) = F(A_1) - F(A_1 \cap A_2) = F(A_2) - F(A_1 \cap A_2) = F(A_2 - A_1)$. Hence the elements in $A_1 - A_2$ add up to the same sum as the elements in $A_2 - A_1$. But $A_1 - A_2$ and $A_2 - A_1$ are disjoint because $A_1 - A_2$ contains no elements of A_2 and $A_2 - A_1$ contains no elements of A_1.

36. <u>Proof:</u> Suppose that 101 integers are chosen from 1 to 200 inclusive. Call them $x_1, x_2, \ldots, x_{101}$. Represent each of these integers in the form $x_i = 2^{k_i} a_i$ where a_i is the uniquely determined odd integer obtained by dividing x_i by the highest possible power of 2. Because each x_i satisfies the condition $1 \leq x_i \leq 200$, each a_i satisfies the condition $1 \leq a_i \leq 199$. Define a function F from $X = \{x_1, x_2, \ldots, x_{101}\}$ to the set Y of all odd integers from 1 to 199 inclusive by the rule $F(x_i) =$ that odd integer a_i such that x_i equals $2^{k_i} \cdot a_i$. Now X has 101 elements and Y has 100 elements, namely

$$1 = 2 \cdot \underline{1} - 1, \; 3 = 2 \cdot \underline{2} - 1, \; 5 = 2 \cdot \underline{3} - 1, \ldots, \; 199 = 2 \cdot \underline{100} - 1.$$

Hence by the pigeonhole principle, F is not one-to-one: there exist integers x_i and x_j such that $F(x_i) = F(x_j)$ and $x_i \neq x_j$.

But $x_i = 2^{k_i} \cdot a_i$ and $x_j = 2^{k_j} \cdot a_j$ and $F(x_i) = a_i$ and $F(x_j) = a_j$. Thus $x_i = 2^{k_i} \cdot a_i$ and $x_j = 2^{k_j} \cdot a_i$. If $k_j > k_i$, then

$$x_j = 2^{k_j} \cdot a_i = 2^{k_j - k_i} \cdot 2^{k_i} \cdot a_i = 2^{k_j - k_i} \cdot x_i,$$

and so x_j is divisible by x_i. Similarly, if $k_j < k_i$, x_i is divisible by x_j. Hence, in either case, one of the numbers is divisible by another.

39. Let S be any set consisting entirely of integers from 1 through 100, and suppose that no integer in S divides any other integer in S. Factor out the highest power of 2 to write each integer in S as $2^k \cdot m$, where m is an odd integer.

    Now consider any two such integers in S, say $2^r \cdot a$ and $2^s \cdot b$. Observe that $a \neq b$. The reason is that if $a = b$, then whichever integer contains the fewer number of factors of 2 divides the other integer. (For example, $2^2 \cdot 3 \mid 2^4 \cdot 3$.)

    Thus there can be no more integers in S than there are distinct odd integers from 1 through 100, namely 50.

    Furthermore, it is possible to find a set T of 50 integers from 1 through 100 no one of which divides any other. For instance, $T = 51, 52, 53, \ldots, 99, 100$.

    Hence the largest number of elements that a set of integers from 1 through 100 can have so that no one element in the set is divisible by any other is 50.

Section 9.5

9. *a.* The number of committees of six that can be formed from the 40 members of the club is

$$\binom{40}{6} = 3,838,380.$$

12. The sum of two integers is even if, and only if, either both integers are even or both are odd *[see Example 4.2.3]*. Because $2 = 2 \cdot 1$ and $100 = 2 \cdot 50$, there are 50 even integers and thus 51 odd integers from 1 to 101 inclusive. Hence the number of distinct pairs is the number of ways to choose two even integers from the 50 plus the number of ways to choose two odd integers from the 51:

$$\binom{50}{2} + \binom{51}{2} = 1225 + 1275 = 2500.$$

18. An ordering for the letters in *MISSISSIPPI* can be created as follows:

    Step 1: Choose a subset of one position for the M

    Step 2: Choose a subset of four positions for the I's

    Step 3: Choose a subset of four positions for the S's

    Step 4: Choose a subset of two positions for the P's

    Thus the total number of distinguishable orderings is

$$\binom{11}{1}\binom{10}{4}\binom{6}{4}\binom{2}{2} = \frac{11!}{1! \cdot 10!} \cdot \frac{10!}{4! \cdot 6!} \cdot \frac{6!}{4! \cdot 2!} \cdot \frac{2!}{2! \cdot 0!} = \frac{11!}{1! \cdot 4! \cdot 4! \cdot 2!} = 34,650,$$

    which agrees with the result in Example 9.5.10.

21. The number of symbols that can be represented in the Morse code using n dots and dashes is 2^n. Therefore, the number of symbols that can be represented in the Morse code using at most seven dots and dashes is

$$2 + 2^2 + 2^3 + 2^4 + 2^5 + 2^6 + 2^7 = 2(1 + 2 + 2^2 + 2^3 + 2^4 + 2^5 + 2^6) = 2\left(\frac{2^7 - 1}{2 - 1}\right) = 254.$$

24. *a.* Because $210 = 2 \cdot 3 \cdot 5 \cdot 7$, the distinct factorizations of 210 are $1 \cdot 210$, $2 \cdot 105$, $3 \cdot 70$, $5 \cdot 42$, $7 \cdot 30$, $6 \cdot 35$, $10 \cdot 21$, and $14 \cdot 15$. So there are 8 distinct factorizations of 210.

c. As in the answer to part (b), there are two different ways to look at the solution to this problem.

Solution 1: Separate the factorizations into categories: one category consists only of the factorization in which one factor is 1 and the other factor is the product of all five prime factors *[there is $1 = \binom{5}{0}$ such factorization]*, a second category consists of those factorizations in which one factor is a single prime and the other factor is the product of the four other primes *[there are $\binom{5}{1}$ such factorizations]*, and the third category contains those factorizations in which one factor is a product of two of the primes and the other factor is the product of the other three primes *[there are $\binom{5}{2}$ such factorizations]*. All possible factorizations are included among these categories, and so, by the addition rule, the answer is $\binom{5}{0} + \binom{5}{1} + \binom{5}{2}$ $= 1 + 5 + 10 = 16$.

Solution 2: Let $S = \{p_1, p_2, p_3, p_4, p_5\}$, let $p_1 p_2 p_3 p_4 p_5 = P$, and let $f_1 f_2$ be any factorization of P. The product of the numbers in any subset $A \subseteq S$ can be used for f_1, with the product of the numbers in A^c being f_2.. Thus there are as many ways to write $f_1 f_2$ as there are subsets of S, namely $2^5 = 32$ (by Theorem 6.3.1). But given any factors f_1 and f_2, we have that $f_1 f_2 = f_2 f_1$. Thus counting the number of ways to write $f_1 f_2$ counts each factorization twice. So the answer is $\frac{32}{2} = 16$.

Note: In Section 9.7 we will show that $\binom{n}{r} = \binom{n}{n-r}$ whenever $n \geq r \geq 0$. Thus, for example, the answer can be written as

$$\binom{5}{0} + \binom{5}{1} + \binom{5}{2} = \frac{1}{2}\left(\binom{5}{0} + \binom{5}{1} + \binom{5}{2} + \binom{5}{3} + \binom{5}{4} + \binom{5}{5}\right).$$

In Section 9.7 we will also show that for all integers $n \geq 0$,

$$\binom{n}{0} + \binom{n}{1} + \binom{n}{2} + \cdots + \binom{n}{n-2} + \binom{n}{n-1} + \binom{n}{n} = 2^n,$$

and so, in particular,

$$\frac{1}{2}\left[\binom{5}{0} + \binom{5}{1} + \binom{5}{2} + \binom{5}{3} + \binom{5}{4} + \binom{5}{5}\right] = \frac{1}{2} \cdot 2^5 = \frac{32}{2} = 16.$$

These facts illustrate the relationship between the two solutions to part (c) of this exercise.

d. Because the second solution given in parts (b) and (c) is the simplest, we give a general version of it as the answer to this part of the exercise. Let $S = \{p_1, p_2, p_3, \ldots, p_n\}$, let $p_1 p_2 p_3 \cdots p_n = P$, and let $f_1 f_2$ be any factorization of P. The product of the numbers in any subset $A \subseteq S$ can be used for f_1, with the product of the numbers in A^c being f_2. Thus there are as many ways to write $f_1 f_2$ as there are subsets of S, namely 2^n (by Theorem 6.3.1). But given any factors f_1 and f_2, we have that $f_1 f_2 = f_2 f_1$, and so counting the number of ways to write $f_1 f_2$ counts each factorization twice. Hence the answer is $\frac{1}{2^n} = 2^{n-1}$.

27. *b.* A reflexive relation must contain (a, a) for all eight elements a in A. Any subset of the remaining 56 elements of $A \times A$ (which has a total of 64 elements) can be combined with these eight to produce a reflexive relation. Therefore, there are as many reflexive binary relations as there are subsets of a set of 56 elements, namely 2^{56}.

d. Form a relation that is both reflexive and symmetric by a two-step process: (1) pick all eight elements of the form (x, x) where $x \in A$, (2) pick a set of (distinct) pairs of elements of the form (a, b) and (b, a). There is just one way to perform step 1, and, as explained in the answer to part (c), there are 2^{28} ways to perform step 2. Therefore, there are 2^{28} binary relations on A that are reflexive and symmetric.

30. The error is that the "solution" overcounts the number of poker hands with two pairs. In fact, it counts every such hand twice. For instance, consider the poker hand $\{4\clubsuit, 4\diamondsuit, J\heartsuit, J\spadesuit, 9\clubsuit\}$. If the steps outlined in the false solution in the exercise statement are followed, this hand is first counted when the denomination 4 is chosen in step one, the cards $4\clubsuit$ and $4\diamondsuit$ are chosen in step two, the denomination J is chosen in step three, the cards J $\heartsuit$ and J$\spadesuit$ are chosen in step four, and $9\clubsuit$ is chosen in step five. The hand is counted a second time when the denomination J is chosen in step one, the cards J$\heartsuit$ and J $\spadesuit$ are chosen in step two, the denomination 4 is chosen in step three, the cards $4\clubsuit$ and $4\diamondsuit$ are chosen in step four, and $9\clubsuit$ is chosen in step five.

Section 9.6

6. $\dbinom{5+n-1}{5} = \dbinom{n+4}{5} = \dfrac{(n+4)(n+3)(n+2)(n+1)n}{120}$

9. The number of iterations of the inner loop is the same as the number of integer triples (i, j, k) where $1 \le k \le j \le i \le n$. As in Example 9.6.3, such triples can be represented as a string of $n-1$ vertical bars and three crosses indicating which three integers from 1 to n are included in the triple. Thus the number of such triples is the same as the number of strings of $(n-1)$ |'s and 3 ×'s, which is

$$\binom{n+2}{3} = \frac{n(n+1)(n+2)}{6}.$$

12. Think of the number 30 as divided into 30 individual units and the variables (y_1, y_2, y_3, y_4) as four categories into which these units are placed. The number of units in category y_i indicates the value of y_i in a solution of the equation. By Theorem 9.6.1, the number of ways to place 30 objects into four categories is

$$\binom{30+4-1}{30} = \binom{33}{30} = 5456.$$

So there are 5456 nonnegative integral solutions of the equation.

15. Any number from 1 through 99,999 whose digits add up to 9 can be thought of as a 5-digit number with leading zeroes included. Imagine that the 5 digits are categories into which we place 9 crosses. (For instance, $\times\times\,|\quad\quad|\times\times\times\times\times\,|\times\,|\times\times$ corresponds to the number 20512.) By Theorem 9.6.1, there are $\binom{9+5-1}{9} = \binom{13}{9} = 715$ ways to place the crosses into the categories.

18. *a.* Think of the 4 kinds of coins as the n categories and the 30 coins to be chosen as the r objects. Each choice of 30 coins is represented by a string of $4-1 = 3$ vertical bars (to separate the categories) and 30 crosses (to represent the chosen coins). The total number of choices of 30 coins of the 4 different kinds is the number of strings of 33 symbols (3 vertical bars and 30 crosses), namely, $\dbinom{30+4-1}{30} = \dbinom{33}{30} = 5,456$.

b. Let T be the set of selections of 30 coins for which the coin's type is unrestricted, $Q_{\le 15}$ the set of selections containing at most 15 quarters, and $Q_{\ge 16}$ the set of selections containing at least 16 quarters. Then

$$T = Q_{\le 15} \cup Q_{\ge 16} \quad \text{and} \quad Q_{\le 15} \cap Q_{\ge 16} = \emptyset \quad \text{and so} \quad N(T) = N(Q_{\le 15}) + N(Q_{\ge 16}).$$

To compute $N(Q_{\ge 16})$, we reason as follows: If at least 16 quarters are included, we can choose the 30 coins by first selecting 16 quarters and then choosing the remaining 14 coins from the four different types. The number of ways to do this is

$$N(Q_{\ge 16}) = \binom{14+4-1}{14} = \binom{17}{14} = 680.$$

Then $N(T) = 5,456$ *[by part (a)]* and $N(Q_{\geq 16}) = 680$. Therefore, the number of selections containing at most 15 quarters is

$$N(Q_{\leq 15}) = N(T) - N(Q_{\geq 16}) = 5,456 - 680 = 4,776.$$

c. Let T be the set of selections of 30 coins for which the coin's type is unrestricted, $D_{\leq 20}$ the set of selections containing at most 20 dimes, and $D_{\geq 21}$ the set of selections containing at least 21 dimes. Then

$$T = D_{\leq 20} \cup D_{\geq 21} \quad \text{and} \quad D_{\leq 20} \cap D_{\geq 21} = \emptyset \quad \text{and so} \quad N(T) = N(D_{\leq 20}) + N(D_{\geq 21}).$$

To compute $N(D_{\geq 21})$, we reason as follows: If at least 21 dimes are included, we can choose the 30 coins by first selecting 21 dimes and then choosing the remaining nine coins from the four different types. The number of ways to do this is

$$N(D_{\geq 21}) = \binom{9 + 4 - 1}{9} = \binom{12}{9} = 220.$$

Then $N(T) = 5,456$ *[by part (a)]* and $N(D_{\geq 21}) = 220$. Therefore, the number of selections containing at most 20 dimes is

$$N(D_{\leq 20}) = N(T) - N(D_{\geq 21}) = 5,456 - 220 = 5,236.$$

d. As in parts (b) and (c), let T be the set of selections of 30 coins for which the coin's type is unrestricted, $Q_{\geq 16}$ the set of selections containing at least 16 quarters, $Q_{\leq 15}$ the set of selections containing at most 15 quarters, $D_{\geq 21}$ the set of selections containing at least 21 dimes, and $D_{\leq 20}$ the set of selections containing at most 20 dimes. If the pile has at most 15 quarters and at most 20 dimes, then the number of combinations of coins that can be chosen is $N(Q_{\leq 15} \cap D_{\leq 20})$, and, by the difference rule,

$$N(Q_{\leq 15} \cap D_{\leq 20}) = N(T) - N(Q_{\geq 16} \cup D_{\geq 21}).$$

In order to find $N(Q_{\geq 16} \cup D_{\geq 21})$, we first compute $N(Q_{\geq 16} \cap D_{\geq 21})$, which is the number of selections of coins containing at least 16 quarters and at least 21 dimes. However, 16 quarters plus 21 dimes would give a total of more than 30 coins. So there are no selections of this type. Thus

$$N(Q_{\leq 15} \cap D_{\leq 20}) = N(T) - N(Q_{\geq 16} \cup D_{\geq 21}) = 5,456 - 0 = 5,456.$$

Then, by the inclusion/exclusion rule,

$$N(Q_{\geq 16} \cup D_{\geq 21}) = N(Q_{\geq 16}) + N(D_{\geq 21}) - N(Q_{\geq 16} \cap D_{\geq 21}) = 680 + 220 - 0 = 900.$$

Therefore the answer to the question is

$$N(Q_{\leq 15} \cap D_{\leq 20}) = N(T) - N(Q_{\geq 16} \cup D_{\geq 21}) = 5,456 - 900 = 4,556.$$

21. Consider those columns of a trace table corresponding to an arbitrary value of k. The values of j go from 1 to k, and for each value of j, the values of i go from 1 to j.

k	k											
j	1	2		3			. . .	k				
i	1	1	2	1	2	3	. . .	1	2	3	. . .	k

So for each value of k, there are $1 + 2 + 3 + \cdots + k$ columns of the table. Since k goes from 1 to n, the total number of columns in the table is

$$1 + (1 + 2) + (1 + 2 + 3) + \cdots + (1 + 2 + 3 + \cdots + n)$$

$$= \sum_{k=1}^{1} k + \sum_{k=1}^{2} k + \cdots + \sum_{k=1}^{n-1} k + \sum_{k=1}^{n} k$$

$$= \frac{1 \cdot 2}{2} + \frac{2 \cdot 3}{2} + \cdots + \frac{(n-1) \cdot n}{2} + \frac{n \cdot (n+1)}{2}$$

$$= \frac{1}{2}[1 \cdot 2 + 2 \cdot 3 + \cdots + (n-1) \cdot n + n \cdot (n+1)]$$

$$= \frac{1}{2}\left(\frac{n(n+1)(n+2)}{3}\right) \qquad \text{by exercise 13 of Section 5.2}$$

$$= \frac{n(n+1)(n+2)}{6},$$

which agrees with the result of Example 9.6.4.

Section 9.7

9.
$$\binom{2(n+1)}{2n} = \binom{2n+2}{2n}$$

$$= \frac{(2n+2)!}{(2n)!((2n+2)-2n)!}$$

$$= \frac{(2n+2)(2n+1)(2n)!}{(2n)!2!}$$

$$= \frac{(2n+2)(2n+1)}{2}$$

$$= \frac{2(n+1)(2n+1)}{2}$$

$$= (n+1)(2n+1)$$

12.
$$\binom{n+3}{r} = \binom{n+2}{r-1} + \binom{n+2}{r}$$

$$= \left(\binom{n+1}{r-2} + \binom{n+1}{r-1}\right) + \left(\binom{n+1}{r-1} + \binom{n+1}{r}\right)$$

$$= \binom{n+1}{r-2} + 2 \cdot \binom{n+1}{r-1} + \binom{n+1}{r}$$

$$= \left(\binom{n}{r-3} + \binom{n}{r-2}\right) + 2 \cdot \left(\binom{n}{r-2} + \binom{n}{r-1}\right) + \left(\binom{n}{r-1} + \binom{n}{r}\right)$$

$$= \binom{n}{r-3} + 3 \cdot \binom{n}{r-2} + 3 \cdot \binom{n}{r-1} + \binom{n}{r}$$

15. Proof (by mathematical induction): Let r be a fixed nonnegative integer, and let the property $P(n)$ be the formula

$$\sum_{i=r}^{n} \binom{i}{r} = \binom{n+1}{r+1}. \qquad \leftarrow Pn)$$

Show that $P(r)$ is true: To prove that $P(r)$ is true, we must show that

$$\sum_{i=r}^{r} \binom{i}{r} = \binom{r+1}{r+1}.$$

But the left-hand side of this equation is $\binom{r}{r} = 1$, and the right-hand side is $\binom{r+1}{r+1}$, which also equals 1. So $P(r)$ is true.

Show that for all integers $k \geq r$, if $P(k)$ is true then $P(k+1)$ is true: Let k be any integer with $k \geq r$ and suppose that

$$\sum_{i=r}^{k} \binom{i}{r} = \binom{k+1}{r+1}. \quad \leftarrow \quad \begin{array}{l} P(k) \\ \text{inductive hypothesis} \end{array}$$

We must show that

$$\sum_{i=r}^{k+1} \binom{i}{r} = \binom{(k+1)+1}{r+1} \quad \leftarrow \quad P(k+1)$$

The left-hand side of $P(k+1)$ is

$$\begin{aligned}
\sum_{i=r}^{k+1} \binom{i}{r} &= \sum_{i=r}^{k} \binom{i}{r} + \binom{k+1}{r} & \text{by writing the last term separately} \\
&= \binom{k+1}{r+1} + \binom{k+1}{r} & \text{by inductive hypothesis} \\
&= \binom{(k+1)+1}{r+1} & \text{by Pascal's formula,}
\end{aligned}$$

and this is the right-hand side of $P(k+1)$ *[as was to be shown]*.

18. <u>Proof (by mathematical induction)</u>: Let the property $P(n)$ be the equation

$$\binom{m}{0} + \binom{m+1}{1} + \binom{m+2}{2} + \cdots + \binom{m+n}{n} = \binom{m+n+1}{n}. \quad \leftarrow P(n)$$

We will show by mathematical induction that the property is true for all integers $n \geq 0$.

Show that $P(0)$ is true: $P(0)$ is the equation $\binom{m}{0} = \binom{m+0+1}{0} = \binom{m+1}{0}$, is true because by exercise 1 both sides equal 1.

Show that for all integers $k \geq 0$, if $P(k)$ is true then $P(k+1)$ is true: Let k be any integer with $k \geq 0$ and suppose that

$$\binom{m}{0} + \binom{m+1}{1} + \binom{m+2}{2} + \cdots + \binom{m+k}{k} = \binom{m+k+1}{k}. \quad \leftarrow \quad \begin{array}{l} P(k) \\ \text{inductive hypothesis} \end{array}$$

We must show that

$$\binom{m}{0} + \binom{m+1}{1} + \binom{m+2}{2} + \cdots + \binom{m+(k+1)}{k+1} = \binom{m+(k+1)+1}{(k+1)}$$

or, equivalently,

$$\binom{m}{0} + \binom{m+1}{1} + \binom{m+2}{2} + \cdots + \binom{m+k+1}{k+1} = \binom{m+k+2}{k+1}. \quad \leftarrow P(k+1)$$

But

$$\begin{aligned}
\binom{m}{0} + \binom{m+1}{1} + \binom{m+2}{2} + \cdots + \binom{m+k+1}{k+1} &= \binom{m+k+1}{k} + \binom{m+k+1}{k+1} \\
& \quad \text{by inductive hypothesis} \\
&= \binom{m+k+2}{k+1} \\
& \quad \text{by Pascal's formula ($m+k+1$ in} \\
& \quad \text{place of n and $k+1$ in place of r).}
\end{aligned}$$

[This is what was to be shown.]

24. *Solution 1*: $(u^2 - 3v)^4 = \binom{4}{0}(u^2)^4(-3v)^0 + \binom{4}{1}(u^2)^3(-3v)^1 + \binom{4}{2}(u^2)^2(-3v)^2$

$$+ \binom{4}{3}(u^2)^1(-3v)^3 + \binom{4}{4}(u^2)^0(-3v)^4$$

$$= u^8 - 12u^6v + 54u^4v^2 - 108u^2v^3 + 81v^4$$

Solution 2: An alternative solution is to first expand and simplify the expression $(a + b)^4$ and then substitute u^2 in place of a and $(-3v)$ in place of b and further simplify the result. Using this approach, we first apply the binomial theorem with $n = 4$ to obtain

$$(a + b)^4 = \binom{4}{0}a^4b^0 + \binom{4}{1}a^3b^1 + \binom{4}{2}a^2b^2 + \binom{4}{3}a^1b^3 + \binom{4}{4}b^4$$

$$= a^4 + 4a^3b + 6a^2b^2 + 4ab^3 + b^4.$$

Substituting u^2 in place of a and $(-3v)$ in place of b gives

$$(u^2 - 3v)^4 = (u^2 + (-3v))^4 = (u^2)^4 + 4(u^2)^3(-3v) + 6(u^2)^2(-3v)^2 + 4(u^2)(-3v)^3 + (-3v)^4$$

$$= u^8 - 12u^6v + 54u^4v^2 - 108u^2v^3 + 81v^4.$$

27. $\left(x^2 - \dfrac{1}{x}\right)^5$

$= (x^2)^5 + \binom{5}{1}(x^2)^4\left(-\dfrac{1}{x}\right)^1 + \binom{5}{2}(x^2)^3\left(-\dfrac{1}{x}\right)^2 + \binom{5}{3}(x^2)^2\left(-\dfrac{1}{x}\right)^3$

$$+ \binom{5}{4}(x^2)^1\left(-\dfrac{1}{x}\right)^4 + \left(-\dfrac{1}{x}\right)^5$$

$$= x^{10} - 5x^7 + 10x^4 - 10x + \dfrac{5}{x^2} - \dfrac{1}{x^5}$$

30. Term is $\binom{10}{3}(2x)^7 3^3$. Coefficient is $\dfrac{10!}{3! \cdot 7!} \cdot 2^7 \cdot 3^3 = 120 \cdot 128 \cdot 27 = 414,720.$

39. <u>Proof</u>: Let n be an integer with $n \geq 0$. Apply the binomial theorem with $a = 3$ and $b = -1$ to obtain

$2^n = (3 + (-1))^n$

$= \binom{n}{0}3^n(-1)^0 + \binom{n}{1}3^{n-1}(-1)^1 + \cdots + \binom{n}{i}3^{n-i}(-1)^i + \cdots + \binom{n}{n}3^{n-n}(-1)^n$

$= \displaystyle\sum_{i=0}^{n}(-1)^i\binom{n}{i}3^{n-i}.$

42. <u>Proof (by mathematical induction)</u>: Let the property $P(n)$ be the sentence

For any set S with n elements, S has 2^{n-1} subsets with an even number of elements and 2^{n-1} subsets with an odd number of elements. $\leftarrow P(n)$

We will prove by mathematical induction that the property is true for all integers $n \geq 1$.

Show that $P(1)$ is true: $P(1)$ is true because any set S with just 1 element, say x, has two subsets: $\emptyset$, which has 0 elements, and $\{x\}$, which has 1 element. Since 0 is even and 1 is odd, the number of subsets of S with an even number of elements equals the number of subsets of S with an odd number of elements, namely, 1, and $1 = 2^0 = 2^{1-1}$.

Show that for all integers $k \geq 1$, if $P(k)$ is true then $P(k+1)$ is true: Let k be any integer with $k \geq 1$ and suppose that

For any set S with k elements, S has 2^{k-1} subsets with an even number of elements and 2^{k-1} subsets with an odd number of elements.

$P(k)$
← inductive hypothesis

We must show that

For any set S with $k+1$ elements, S has $2^{(k+1)-1}$ subsets with an even number of elements and $2^{(k+1)-1}$ subsets with an odd number of elements.

or, equivalently,

For any set S with $k+1$ elements, S has 2^k subsets with an even number of elements and 2^k subsets with an odd number of elements.

← $P(k+1)$

Call the elements of $S = \{x_1, x_2, \ldots, x_k, x_{k+1}\}$. By inductive hypothesis, $\{x_1, x_2, \ldots, x_k\}$ has 2^{k-1} subsets with an even number of elements and 2^{k-1} subsets with an odd number of elements. Now every subset of $\{x_1, x_2, \ldots, x_k\}$ is also a subset of S, and the only other subsets of S are obtained by taking the union of a subset of $\{x_1, x_2, \ldots, x_k\}$ with $\{x_{k+1}\}$. Moreover, if a subset of $\{x_1, x_2, \ldots, x_k\}$ has an even number of elements, then the union of that subset with $\{x_{k+1}\}$ has an odd number of elements. So 2^{k-1} of the subsets of S that are obtained by taking the union of a subset of $\{x_1, x_2, \ldots, x_k\}$ with $\{x_{k+1}\}$ have an even number of elements and 2^{k-1} have an odd number of elements. Thus the total number of subsets of S with an even number of elements is

$$2^{k-1} + 2^{k-1} = 2 \cdot 2^{k-1} = 2^{1+(k-1)} = 2^k.$$

Similarly, the total number of subsets of S with an odd number of elements is also

$$2^{k-1} + 2^{k-1} = 2^k$$

[as was to be shown].

Alternative justification for the identity in exercise 36: Let n be any positive integer, let E be the largest even integer less than or equal to n, and let O be the largest odd integer less than or equal to n. Let S be any set with n elements. Then the number of subsets of S with an even number of elements is $\binom{n}{0} + \binom{n}{2} + \binom{n}{4} + \cdots + \binom{n}{E}$, and the number of subsets of S with an odd number of elements is $\binom{n}{1} + \binom{n}{3} + \binom{n}{5} + \cdots + \binom{n}{O}$. But there are as many subsets with an even number of elements as there are subsets with an odd number of elements, so if we subtract the second of these quantities from the first we obtain 0:

$$\begin{aligned} 0 &= \left[\binom{n}{0} + \binom{n}{2} + \binom{n}{4} + \cdots + \binom{n}{E}\right] - \left[\binom{n}{1} + \binom{n}{3} + \binom{n}{5} + \cdots + \binom{n}{O}\right] \\ &= \binom{n}{0} - \binom{n}{1} + \binom{n}{2} - \binom{n}{3} + \cdots + (-1)^n \binom{n}{n}. \end{aligned}$$

48. Let n be an integer with $n \geq 0$. Then

$$\begin{aligned} \sum_{r=0}^{n} \binom{n}{r} x^{2r} &= \sum_{r=0}^{n} \binom{n}{r} 1^{n-r} (x^2)^r \\ & \qquad \text{by the laws of exponents and because 1 raised to any power is 1} \\ &= (1+x^2)^n \\ & \qquad \text{by the binomial theorem with } a = 1 \text{ and } b = x^2. \end{aligned}$$

52. Let n be an integer with $n \geq 0$. Then

$$\sum_{k=0}^{n} \binom{n}{k} 3^{2n-2k} 2^{2k} = \sum_{k=0}^{n} \binom{n}{k} (3^2)^{n-k} (2^2)^k$$

by the laws of exponents

$$= \sum_{k=0}^{n} \binom{n}{k} 9^{n-k} 4^k$$

because $3^2 = 9$ and $2^2 = 4$

$$= (9+4)^n$$

by the binomial theorem with $a = 9$ and $b = 4$

$$= 13^n.$$

Section 9.8

3. *a.* $P(A \cup B) = 0.4 + 0.2 = 0.6$

   b. By the formula for the probability of a general union and because $S = A \cup B \cup C$,

   $$P(S) = ((A \cup B) \cup C) = P(A \cup B) + P(C) - P((A \cup B) \cap C).$$

   Suppose $P(C) = 0.2$. Then, since $P(S) = 1$,

   $$1 = 0.6 + 0.2 - P((A \cup B) \cap C) = 0.8 - P((A \cup B) \cap C).$$

   Solving for $P((A \cup B) \cap C)$ gives $P((A \cup B) \cap C) = -0.2$, which is impossible. Hence $P(C) \neq 0.2$.

6. First note that we can apply the formula for the probability of the complement of an event to obtain $0.3 = P(U^c) = 1 - P(U)$. Solving for $P(U)$ gives $P(U) = 0.7$. Second, observe that by De Morgan's law $U^c \cup V^c = (U \cap V)^c$. Thus

   $$0.4 = P(U^c \cup V^c) = P((U \cap V)^c) = 1 - P(U \cap V).$$

   Solving for $P(U \cap V)$ gives $P(U \cap V) = 0.6$. So, by the formula for the union of two events,

   $$P(U \cup V) = P(U) + P(V) - P(U \cap V) = 0.7 + 0.6 - 0.6 = 0.7.$$

9. *b.* By part (a), $P(A \cup B) = 0.7$. So, since $C = (A \cup B)^c$, by the formula for the probability of the complement of an event,

   $$P(C) = 1 - P(A \cup B) = 1 - 0.7 = 0.3.$$

   c. By the formula for the probability of the complement of an event,

   $$P(A^c) = 1 - P(A) = 1 - 0.4 = 0.6.$$

   e. By De Morgan's law $A^c \cup B^c = (A \cap B)^c$. Thus, the formula for the probability of the complement of an event,

   $$P(A^c \cup B^c) = P((A \cap B)^c) = 1 - P(A \cap B) = 1 - 0.2 = 0.8.$$

   f. Solution 1: Because $C = S - (A \cup B)$, we have that $C = (A \cup B)^c$. Then

$$
\begin{array}{lll}
B^c \cap C &= B^c \cap (A \cup B)^c & \text{by substitution} \\
&= B^c \cap (A^c \cap B^c) & \text{by De Morgan's law} \\
&= (B^c \cap A^c) \cap B^c & \text{by the associative law for } \cap \\
&= (A^c \cap B^c) \cap B^c & \text{by the commutative law for } \cap \\
&= A^c \cap (B^c \cap B^c) & \text{by the associative law for } \cap \\
&= A^c \cap B^c & \text{by the idempotent law for } \cap \\
&= (A \cup B)^c & \text{by De Morgan's law} \\
&= C & \text{by substitution.}
\end{array}
$$

Hence, by part (b), $P(B^c \cap C) = P(C) = 0.3$.

Solution 2: Because $C = S - (A \cup B)$, we have that $C = (A \cup B)^c$. Thus by De Morgan's law, $C = A^c \cap B^c$. Now $A^c \cap B^c \subseteq B^c$ *[by Theorem 9.2.1(1)b]* and hence $B^c \cap C = C$ *[by Theorem 9.2.3a]*. Therefore $P(B^c \cap C) = P(C) = 0.3$.

12. <u>Proof 1</u>: Suppose S is any sample space and U and V are any events in S. First note that by the set difference, distributive, universal bound, and identity laws,

$$(V \cap U) \cup (V - U) = (V \cap U) \cup (V \cap U^c) = V \cap (U \cup U^c) = V \cap S = V.$$

Next, observe that if $x \in (V \cap U) \cap (V - U)$, then, by definition of intersection, $x \in (V \cap U)$ and $x \in (V - U)$, and so, by definition of intersection and set difference, $x \in V$, $x \in U$, $x \in V$, and $x \notin U$, and hence, in particular, $x \in U$ and $x \notin U$, which is impossible. It follows that $(V \cap U) \cap (V - U) = \emptyset$. Thus, by substitution and by probability axiom 3 (the formula for the probability of mutually disjoint events),

$$P(V) = P((V \cap U) \cup (V - U)) = P(V \cap U) + P(V - U).$$

Solving for $P(V - U)$ gives

$$P(V - U) = P(V) - P(U \cap V).$$

<u>Proof 2</u>: Suppose S is any sample space and U and V are any events in S. First note that by the set difference, distributive, universal bound, and identity laws,

$$U \cup (V - U) = U \cup (V \cap U^c) = (U \cup V) \cap (U \cup U^c) = (U \cup V) \cap S = U \cup V.$$

Also by the set difference law, and the associative, commutative, and universal bound laws for $\cap$,

$$U \cap (V - U) = U \cap (V \cap U^c) = U \cap (U^c \cap V) = (U \cap U^c) \cap V = \emptyset \cap V = \emptyset.$$

Thus, by probability axiom 3 (the formula for the probability of mutually disjoint events),

$$P(U \cup V) = P(U \cup (V - U)) = P(U) + P(V - U).$$

But also by the formula for the probability of a general union,

$$P(U \cup V) = P(U) + P(V) - P(U \cap V).$$

Equating the two expressions for $P(U \cup V)$ gives

$$P(U) + P(V - U) = P(U) + P(V) - P(U \cap V).$$

Subtracting $P(U)$ from both sides gives

$$P(V - U) = P(V) - P(U \cap V).$$

15. *Solution 1*: The net gain for the first prize winner is $\$10,000,000 - \$0.60 = \$9,999,999.40$, that for the second prize winner is $\$1,000,000 - \$0.60 = \$999,999.40$, and that for the third prize winner is $\$50,000 - \$0.60 = \$49,999.40$. Each of the other 29,999,997 million people who mail back an entry form has a net loss of $\$0.60$. Because all of the 30 million entry forms have an equal chance of winning the prizes, the expected gain or loss is

$$\$9999999.40 \cdot \frac{1}{30000000} + \$999999.40 \cdot \frac{1}{30000000} + \$49999.40 \cdot \frac{1}{30000000} - \$0.60 \cdot \frac{29999997}{30000000}) \cong -\$0.23,$$

or an expected loss of about 23 cents per person.

Solution 2: The total amount spent by the 30 million people who return entry forms is $30,000,000 \cdot \$0.60 = \$18,000,000$. The total amount of prize money awarded is $\$10,000,000 + \$1,000,000 + \$50,000 = \$11,050,000$. Thus the net loss is $\$18,000,000 - \$11,050,000 = \$6,950,000$, and so the expected loss per person is $6950000/30000000 \cong -\$0.23$, or about 23 cents per person.

18. Let 2_1 and 2_2 denote the two balls with the number 2, let 8_1 and 8_2 denote the two balls with the number 8, and let 1 denote the other ball. There are $\binom{5}{3} = 10$ subsets of 3 balls that can be chosen from the urn. The following table shows the sums of the numbers on the balls in each set and the corresponding probabilities:

Subset	Sum s	Probability of s
$\{1, 2_1, 2_2\}$	5	1/10
$\{1, 2_1, 8_1\}, \{1, 2_2, 8_1\}, \{1, 2_1, 8_2\}, \{1, 2_2, 8_2\}$	11	4/10
$\{2_1, 2_2, 8_1\}, \{2_1, 2_2, 8_2\}$	12	2/10
$\{1, 8_1, 8_2\}$	17	1/10
$\{2_1, 8_1, 8_2\}, \{2_2, 8_1, 8_2\}$	18	2/10

Thus the expected value is $5 \cdot \dfrac{1}{10} + 11 \cdot \dfrac{4}{10} + 12 \cdot \dfrac{2}{10} + 17 \cdot \dfrac{1}{10} + 18 \cdot \dfrac{2}{10} = \dfrac{126}{10} = 12.6$.

21. When a coin is tossed 4 times, there are $2^4 = 16$ possible outcomes and there are $\binom{4}{h}$ ways to obtain exactly h heads (as shown by the technique illustrated in Example 9.5.9). The following table shows the possible outcomes of the tosses, the amount gained or lost for each outcome, the number of ways the outcomes can occur, and the probabilities of the outcomes.

Number of Heads	Net Gain (or Loss)	Number of Ways	Probability
0	$-\$3$	$\binom{4}{0} = 1$	1/16
1	$-\$2$	$\binom{4}{1} = 4$	4/16
2	$-\$1$	$\binom{4}{2} = 6$	6/16
3	$\$2$	$\binom{4}{3} = 4$	4/16
4	$\$3$	$\binom{4}{4} = 1$	1/16

Thus the expected value is $(-\$3)\cdot\dfrac{1}{16}+(-\$2)\cdot\dfrac{4}{16}+(-\$1)\cdot\dfrac{6}{16}+\$2\cdot\dfrac{4}{16}+\$3\cdot\dfrac{1}{16} = -\$\dfrac{6}{16} = -\$0.375$. So this game has an expected loss of 37.5 cents.

Section 9.9

3. Of the students who received A's on the first test, the percent who also received A's on the second test is

$$\frac{\text{the percent of students who received } A\text{'s on both tests}}{\text{the percent of students who received } A\text{'s on the first test}} = \frac{15\%}{25\%} = 0.6 = 60\%.$$

Thus the probability that a person who has the condition tests positive for it is 99%.

9. Proof: Suppose that a sample space S is a union of two disjoint events B_1 and B_2, that A is an event in S with $P(A) \neq 0$, and that $P(B_k) \neq 0$ for $k = 1$ and $k = 2$. Because B_1 and B_2 are disjoint, the same reasoning as in Example 9.9.5 establishes that

$$A = (A \cap B_1) \cup (A \cap B_2) \quad \text{and} \quad (A \cap B_1) \cap (A \cap B_2) = \emptyset.$$

Thus

$$P(A) = P(A \cap B_1) + P(A \cap B_2).$$

Moreover, for each $k = 1$ or 2, by definition of conditional probability, we have both

$$P(B_k \mid A) = \frac{P(B_k \cap A)}{P(A)} = \frac{P(A \cap B_k)}{P(A)} \quad \text{and} \quad P(A \cap B_k) = P(A \mid B_k)P(B_k).$$

Putting these results together gives that for each $k = 1$ or 2,

$$P(B_k \mid A) = \frac{P(A \cap B_k)}{P(A)} = \frac{P(A \mid B_k)P(B_k)}{P(A \cap B_1) + P(A \cap B_2)} = \frac{P(A \mid B_k)P(B_k)}{P(A \mid B_1)P(B_1) + P(A \mid B_2)P(B_2)},$$

which is Bayes' theorem for $n = 2$.

12. *a.* Let B_1 be the event that the first urn is chosen, B_2 the event that the second urn is chosen, and A the event that the chosen ball is blue. Then

$$P(A \mid B_1) = \frac{4}{20} \quad \text{and} \quad P(A \mid B_2) = \frac{10}{19}.$$

$$P(A \cap B_1) = P(A \mid B_1)P(B_1) = \frac{4}{20} \cdot \frac{1}{2} = \frac{1}{10}.$$

Also

$$P(A \cap B_2) = P(A \mid B_2)P(B_2) = \frac{10}{19} \cdot \frac{1}{2} = \frac{5}{19}.$$

Now A is the disjoint union of $A \cap B_1$ and $A \cap B_2$. So

$$P(A) = P(A \cap B_1) + P(A \cap B_2) = \frac{1}{10} + \frac{5}{19} = \frac{69}{190} \cong 36.3\%.$$

Thus the probability that the chosen ball is blue is approximately 36.3%.

b. Solution 1 (using Bayes' theorem): Given that the chosen ball is blue, the probability that it came from the first urn is $P(B_1 \mid A)$. By Bayes' theorem and the computations in part (a),

$$P(B_1 \mid A) = \frac{P(A \mid B_1)P(B_1)}{P(A \mid B_1)P(B_1) + P(A \mid B_2)P(B_2)} = \frac{\frac{1}{10}}{\frac{1}{10} + \frac{5}{19}} = \frac{19}{69} \cong 27.5$$

Solution 2 (without explicit use of Bayes' theorem): Given that the chosen ball is blue, the probability that it came from the first urn is $P(B_1 \mid A)$. By the results of part (a),

$$P(B_1 \mid A) = \frac{P(A \cap B_1)}{P(A)} = \frac{\frac{1}{10}}{\frac{69}{190}} = \frac{19}{69} \cong 27.5.$$

15. Let B_1 be the event that the part came from the first factory, B_2 the event that the part came from the second factory, and A the event that a part chosen at random from the 180 is defective.

a. The probability that a part chosen at random from the 180 is from the first factory is $P(B_1) = \frac{100}{180}$.

b. The probability that a part chosen at random from the 180 is from the second factory is $P(B_2) = \frac{80}{180}$.

c. The probability that a part chosen at random from the 180 is defective is $P(A)$. Because 2% of the parts from the first factory and 5% of the parts from the second factory are defective, $P(A \mid B_1) = \frac{2}{100}$ and $P(A \mid B_2) = \frac{5}{100}$. By definition of conditional probability,

$$P(A \cap B_1) = P(A \mid B_1)P(B_1) = \frac{2}{100} \cdot \frac{100}{180} = \frac{1}{90}$$

$$P(A \cap B_2) = P(A \mid B_2)P(B_2) = \frac{5}{100} \cdot \frac{80}{180} = \frac{2}{90}.$$

Now because B_1 and B_2 are disjoint and because their union is the entire sample space, A is the disjoint union of $A \cap B_1$ and $A \cap B_2$. Thus the probability that

$$P(A) = P(A \cap B_1) + P(A \cap B_2) = \frac{1}{90} + \frac{2}{90} = \frac{3}{90} \cong 3.3\%.$$

d. Solution 1 (using Bayes' theorem): Given that the chosen part is defective, the probability that it came from the first factory is $P(B_1 \mid A)$. By Bayes' theorem and the computations in part (a),

$$P(B_1 \mid A) = \frac{P(A \mid B_1)P(B_1)}{P(A \mid B_1)P(B_1) + P(A \mid B_2)P(B_2)} = \frac{\frac{1}{90}}{\frac{1}{90} + \frac{2}{90}} = \frac{1}{3} \cong 33.3\%.$$

Solution 2 (without explicit use of Bayes' theorem): Given that the chosen ball is green, the probability that it came from the first urn is $P(B_1 \mid A)$. By the results of part (a),

$$P(B_1 \mid A) = \frac{P(A \cap B_1)}{P(A)} = \frac{\frac{1}{90}}{\frac{3}{90}} = \frac{1}{3} \cong 33.3\%.$$

18. <u>Proof</u>: Suppose A and B are events in a sample space S, and $P(A \cap B) = P(A)\,P(B)$, $P(A) \neq 0$, and $P(B) \neq 0$. Applying the hypothesis to the definition of conditional probability gives

$$P(A \mid B) = \frac{P(A \cap B)}{P(B)} = \frac{P(A)\,P(B)}{P(B)} = P(A)$$

and

$$P(B \mid A) = \frac{P(A \cap B)}{P(A)} = \frac{P(A)\,P(B)}{P(A)} = P(B).$$

21. If A and B are events in a sample space and $A \cap B = \emptyset$ and A and B are independent, then (by definition of independence) $P(A \cap B) = P(A)P(B)$, and (because $A \cap B = \emptyset$) $P(A \cap B) = 0$. Hence $P(A)\,P(B) = 0$, and so (by the zero product property) either $P(A) = 0$ or $P(B) = 0$.

24. Let A be the event that a randomly chosen error is missed by proofreader X, and let B be the event that the error is missed by proofreader Y. Then $P(A) = 0.12$ and $P(B) = 0.15$.

    a. Because the proofreaders work independently, $P(A \cap B) = P(A)\,P(B)$. Hence the probability that the error is missed by both proofreaders is

$$P(A \cap B) = P(A)\,P(B) = (0.12)(0.15) = 0.018 = 1.8\%.$$

    b. Assuming that the manuscript contains 1000 typographical errors, the expected number of missed errors is $1000 \cdot 0.018\% = 18$.

27. *Solution*: The family could have two boys, two girls, or one boy and one girl.

    Let the subscript 1 denote the firstborn child (understanding that in the case of twins this might be by only a few moments), and let the subscript 2 denote the secondborn child.

    Then we can let (B_1G_2, B_1) denote the outcome that the firstborn child is a boy, the secondborn is a girl, and the child you meet is the boy.

    Similarly, we can let (B_1B_2, B_2) denote the outcome that both the firstborn and the secondborn are boys and the child you meet is the secondborn boy.

    When this notational scheme is used for the entire set of possible outcomes for the genders of the children and the gender of the child you meet, all outcomes are equally likely and the sample space is denoted by

    $\{(B_1B_2, B_1), (B_1B_2, B_2), (B_1G_2, B_1), (B_1G_2, G_2), (G_1B_2, G_1), (G_1B_2, B_2), (G_1G_2, G_1), (G_1G_2, G_2)\}.$

    The event that you meet one of the children and it is a boy is

$$\{(B_1B_2, B_1), (B_1B_2, B_2), (B_1G_2, B_1), (G_1B_2, B_2)\}.$$

The probability of this event is $4/8 = 1/2$.

Discussion: An intuitive way to see this conclusion is to realize that the fact that you happen to meet one of the children and see that it is a boy gives you no information about the gender of the other child. Because each of the children is equally likely to be a boy, the probability that the other child is a boy is $1/2$.

Consider the following situation in which the probabilities are identical to the situation described in the exercise. A person tosses two fair coins and immediately covers them so that you cannot see which faces are up. The person then reveals one of the coins, and you see that it is heads. This action on the person's part has given you no information about the other coin; the probability that the other coin has also landed heads up is $1/2$.

30. *a.* $P(0 \text{ false positives}) = \left[\begin{array}{c} \text{the number of ways 0 false} \\ \text{positives can be obtained} \\ \text{over a ten-year period} \end{array} \right] \left(P\left(\begin{array}{c} \text{false} \\ \text{positive} \end{array} \right) \right)^0 \left(P\left(\begin{array}{c} \text{not a false} \\ \text{positive} \end{array} \right) \right)^{10}$

$$= \binom{10}{0} 0.96^{10} = 1 \cdot 0.96^{10} \cong 0.665 = 66.5\%$$

c. $P(2 \text{ false positives}) = \left[\begin{array}{c} \text{the number of ways 2 false} \\ \text{positives can be obtained} \\ \text{over a ten-year period} \end{array} \right] \left(P\left(\begin{array}{c} \text{false} \\ \text{positive} \end{array} \right) \right)^2 \left(P\left(\begin{array}{c} \text{not a false} \\ \text{positive} \end{array} \right) \right)^8$

$$= \binom{10}{2} 0.04^2 \cdot 0.96^8 = 45 \cdot 0.04^2 \cdot 0.96^8 = 0.05194 \cong 5.2\%$$

d. Let T be the event that a woman's test result is positive one year, and let C be the event that the woman has breast cancer.

(i) By Bayes' formula, the probability of C given T is

$$P(C \mid T) = \frac{P(T \mid C)P(C)}{P(T \mid C)P(C) + P(T \mid C^c)P(C^c)}$$

$$= \frac{(0.98)(0.0002)}{(0.98)(0.0002) + (0.04)(0.9998)}$$

$$\cong 0.00488 = 4.88\%.$$

(ii) The event that a woman's test result is negative one year is T^c. By Bayes formula, the probability of C given T^c is

$$P(C \mid T^c) = \frac{P(T^c \mid C)P(C)}{P(T^c \mid C)P(C) + P(T^c \mid C^c)P(C^c)}$$

$$= \frac{(0.02)(0.0002)}{(0.02)(0.0002) + (0.96)(0.9998)}$$

$$\cong 0.000004 = 0.0004\%.$$

33. Suppose a gambler starts with $\$k$. Rolling a fair die leads to one of two disjoint outcomes: winning $\$1$ or losing $\$1$. Let A_k be the event that the gambler is ruined when he has $\$k$. Then A_k is the disjoint union of the following two events: C_k and D_k, where

   C_k is the event that the gambler has $\$k$, wins the next roll, and eventually gets ruined

and D_k is the event that the gambler has $\$k$, loses the next roll, and eventually gets ruined.

Now P_k is the probability that the gambler eventually gets ruined when he has \$$k$. By probability axiom 3,

$$P_k = P(C_k) + P(D_k).$$

Let W be the event that the gambler wins on any given roll. Then

$$P(W) = \frac{1}{6} \quad \text{and} \quad P(W^c) = \frac{5}{6}.$$

For each integer k with $1 \le k \le 300$, the definition of conditional probability can be used to find $P(C_k)$ and $P(D_k)$:

$$
\begin{aligned}
P(C_k) \;=\; & P(A_k \cap W) \\
& \quad \text{by definition of } C_k,\ A_k,\ \text{and } W \\
=\; & P(A_k \,|\, W)P(W) \\
& \quad \text{by definition of conditional probability} \\
=\; & P(A_{k+1}) \cdot \frac{1}{6} \\
& \quad \text{because if the gambler wins on a roll when he has \$}k \\
& \quad \text{then on the next roll he has \$}(k+1) \\
=\; & P_{k+1} \cdot \frac{1}{6}.
\end{aligned}
$$

Similarly,

$$
\begin{aligned}
P(D_k) \;=\; & P(A_k \cap W^c) \\
& \quad \text{by definition of } C_k,\ A_k,\ \text{and } W \\
=\; & P(A_k \,|\, W^c)P(W^c) \\
& \quad \text{by definition of conditional probability} \\
=\; & P(A_{k-1}) \cdot \frac{5}{6} \\
& \quad \text{because if the gambler loses on a roll when he has \$}k \\
& \quad \text{then on the next roll he has \$}(k-1) \\
=\; & P_{k-1} \cdot \frac{5}{6}.
\end{aligned}
$$

Thus,

$$P_k = P(C_k) + P(D_k) = P_{k+1} \cdot \frac{1}{6} + P_{k-1} \cdot \frac{5}{6}.$$

Review Guide: Chapter 9

Probability

- What is the sample space of an experiment? *(p. 518)*
- What is an event in the sample space? *(p. 518)*
- What is the probability of an event when all the outcomes are equally likely? *(p. 518)*

Counting

- If m and n are integers with $m \leq n$, how many integers are there from m to n inclusive? *(p. 521)*
- How do you construct a possibility tree? *(p. 525)*
- What are the multiplication rule, the addition rule, and the difference rule? *(pp. 527, 540, 541)*
- What is the inclusion/exclusion rule? *(p. 545 and exercise 48 on p. 553)*
- What is a permutation? an r-permutation? *(pp. 531, 533)*
- What is $P(n, r)$? *(p. 533)*
- How does the multiplication rule give rise to $P(n, r)$? *(pp. 533-534)*
- When should you use the multiplication rule and when should you use the addition rule? *(p. 577)*
- What are some situations where both the multiplication and the addition or difference rule must be used? *(pp. 540-545)*
- What is the formula for the probability of the complement of an event? *(p. 543)*
- How are IP addresses created? *(p. 544)*
- How is the inclusion/exclusion rule used? *(pp. 546-549)*
- What is an r-combination? *(p. 566)*
- What is an unordered selection of elements from a set? *(p. 566)*
- What is complete enumeration? *(p. 567)*
- What formulas are used to compute $\binom{n}{r}$ by hand? *(p. 568)*
- What are some situations where both r-combinations and the addition or difference rule must be used? *(pp. 569-571)*
- What are some situations where r-combinations, the multiplication rule, and the addition rule are all needed? ? *(pp. 573-574)*
- How can r-combinations be used to count the number of permutations of a set with repeated elements? *(pp. 575-576)*
- What are some formulas for the number of permutations of a set of objects when some of the objects are indistinguishable from each other? *(p. 577)*
- What are Stirling numbers of the second kind? How do you find a recurrence relation for the number of ways a set of size n can be partitioned into r subsets? *(pp. 578-580)*
- What is an r-combination with repetition allowed (or a multiset of size r)? *(p. 584)*
- How many r-combinations with repetition allowed can be selected from a set of n elements? *(p. 586)*

The Pigeonhole Principle

- What is the pigeonhole principle? *(p. 554)*
- How is the pigeonhole principle used to show that rational numbers have terminating or repeating decimal expansions? *(pp. 557-559)*

- What is the generalized pigeonhole principle? *(p. 559)*
- What is the relation between one-to-one and onto for a function defined from one finite set to another of the same size? *(p. 562)*

Pascal's Formula and the Binomial Theorem

- What is Pascal's formula? Can you apply it in various situations? *(p. 593)*
- What is the algebraic proof of Pascal's formula? *(p. 595)*
- What is the combinatorial proof of Pascal's formula? *(pp. 595-596)*
- What is the binomial theorem? Can you apply it in various situations? *(p. 598)*
- What is the algebraic proof of the binomial theorem? *(p. 598-600)*
- What is the combinatorial proof of the binomial theorem? *(pp. 600-601)*

Probability Axioms and Expected Value

- What is the range of values for the probability of an event? *(p. 605)*
- What is the probability of an entire sample space? *(p. 605)*
- What is the probability of the empty set? *(p. 605)*
- If A and B are disjoint events in a sample space S, what is $P(A \cup B)$? *(p. 605)*
- If A is an event in a sample space S, what is $P(A^c)$? *(p. 605)*
- If A and B are any events in a sample space S, what is $P(A \cup B)$? *(p. 606)*
- How do you compute the expected value of a random experiment or process, if the possible outcomes are all real numbers and you know the probability of each outcome? *(p. 608)*
- What is the conditional probability of one event given another event? *(p. 612)*
- What is Bayes' theorem? *(p. 616)*
- What does it mean for two events to be independent? *(p. 618)*
- What is the probability of an intersection of two independent events? *(p. 618)*
- What does it mean for events to be mutually independent? *(p. 620)*
- What is the probability of an intersection of mutually independent events? *(p. 621)*

Chapter 10: Graphs and Trees

The first section of this chapter introduces the terminology of graph theory, illustrating it in a variety of different instances. Several exercises are designed to clarify the distinction between a graph and a drawing of a graph. You might point out to students the advantage of the formal definition over the informal drawing for computer representation of graphs. Other exercises explore the use of graphs to solve problems of various sorts. In some cases, students may be able to solve the given problems, such as the wolf, the goat, the cabbage and the ferryman, more easily without using graphs than using them. The point to make is that such problems *can* be solved using graphs and that for more complex problems involving, say, hundreds of possible states, a graphical representation coupled with a computer path-finding algorithm makes it possible find a solution that could not be discovered by trial-and-error alone. The variety of solutions for exercise 33, on the number of edges of a complete graph illustrates the relations among different branches of discrete mathematics. The rest of the exercises in this section are intended to give you practice in applying the theorem that relates the total degree of a graph to the number of its edges, especially for exploring properties of simple graphs, complete graphs, and bipartite graphs.

In Section 10.2 the general topic of trails, paths and circuits is discussed, including the notion of connectedness and Euler and Hamiltonian circuits. As in the rest of the chapter, an attempt is made to balance the presentation of theory and application.

Section 10.3 introduces the concept of the adjacency matrix of a graph. The main theorem of the section states that the ijth entry of the kth power of the adjacency matrix equals the number of walks of length k from the ith to the jth vertices in the graph. Matrix multiplication is defined and explored in this section in a way that is intended to be adequate even if you have never seen the definition before.

The concept of graph isomorphism is discussed in Section 10.4. In this section the main theorem gives a list of isomorphic invariants that can be used to determine the non-isomorphism of two graphs.

The last three sections of the chapter deal with the subject of trees. Section 10.5 focuses on basic definitions, examples, and theorems giving necessary and sufficient conditions for graphs to be trees, and Section 10.6 contains the definition of rooted tree, binary tree, and the theorems that relate the number of internal to the number of terminal vertices of a full binary tree and the maximum height of a binary tree to the number of its terminal vertices. Section 10.7 on spanning trees and shortest paths contains Kruskal's, Prim's, and Dijkstra's algorithms and proofs of their correctness, as well as applications of minimum spanning trees and shortest paths.

Section 10.1

6.

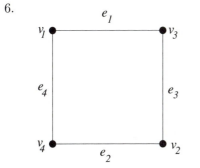

9. (i) e_1, e_2, e_7 are incident on e_1.

(ii) v_1 and v_2 are adjacent to v_3.

(iii) e_2 and e_7 are adjacent to e_1.

(iv) e_1 and e_3 are loops.

(v) e_4 and e_5 are parallel.

(vi) v_4 is an isolated vertex.

(vii) degree of $v_3 = 2$

(viii) total degree of the graph $= 14$

27. *b.* Yes. Each could be friends with all three others.

30. Let t be the total degree of the graph. Since the degree of each vertex is at least d_{min} and at most d_{max}, $d_{min} \cdot v \leq t \leq d_{max} \cdot v$. But by Theorem 10.1.1, t equals twice the number of edges. So by substitution, $d_{min} \cdot v \leq 2e \leq d_{max} \cdot v$. Dividing each part of the inequality by 2 produces the required result:

$$\frac{1}{2} d_{min} \cdot v \;\leq\; e \;\leq\; \frac{1}{2} d_{max} \cdot v.$$

33. *b.* <u>Proof 1</u>: Suppose n is an integer with $n \geq 1$ and K_n is a complete graph on n vertices. If $n = 1$, then K_n has one vertex and 0 edges and $\frac{n(n-1)}{2} = \frac{1(1-1)}{2} = 0$, and so K_n has $\frac{n(n-1)}{2}$ edges. If $n \geq 2$, then since each pair of distinct vertices of K_n is connected by exactly one edge, there are as many edges in K_n as there are subsets of size two of the set of n vertices. By Theorem 9.5.1, there are $\binom{n}{2}$ such sets. But

$$\binom{n}{2} = \frac{n!}{2!(n-2)!} = \frac{n(n-1)}{2}.$$

Hence there are $\frac{n(n-1)}{2}$ edges in K_n.

<u>Proof 2 (by mathematical induction</u>: Let the property $P(n)$ be the sentence

$$\text{A complete graph on } n \text{ vertices has } \frac{n(n-1)}{2} \text{ edges.} \qquad \leftarrow P(n)$$

We will prove that $P(n)$ is true for all integers $n \geq 1$.

Show that $P(1)$ ***is true***: $P(1)$ is true because a complete graph on one vertex has 0 edges and the quantity $\frac{n(n-1)}{2} = \frac{1(1-1)}{2} = 0$ also.

Show that for all integers $m \geq 1$***, if*** $P(m)$ ***is true then*** $P(m+1)$ ***is true***: Let m be any integer with $m \geq 1$, and suppose

$$\text{A complete graph on } m \text{ vertices has } \frac{m(m-1)}{2} \text{ edges.} \qquad \leftarrow \begin{array}{l} P(m) \\ \text{inductive hypothesis} \end{array}$$

We must show that

$$\text{A complete graph on } m+1 \text{ vertices has } \frac{(m+1)((m+1)-1)}{2} \text{ edges,}$$

or, equivalently,

$$\text{A complete graph on } m+1 \text{ vertices has } \frac{m(m+1)}{2} \text{ edges.} \qquad \leftarrow P(m+1)$$

Let K_{m+1} be a complete graph on $m+1$ vertices. Temporarily remove one vertex, v, together with all the edges joining this vertex to the other vertices of the graph. In the graph thus obtained, each vertex is connected to each other vertex by exactly one edge, and so the graph is a complete graph on m vertices. By inductive hypothesis this graph has $\frac{m(m-1)}{2}$ edges. Connecting v to each of the m other vertices adds another m edges. Hence the total number of edges of K_{m+1} is

$$\frac{m(m-1)}{2} + m = \frac{m(m-1)}{2} + \frac{2m}{2} = \frac{m^2 - m + 2m}{2} = \frac{m(m+1)}{2}$$

[as was to be shown].

<u>Proof 3</u>: Suppose n is an integer with $n \geq 1$ and K_n is a complete graph on n vertices. Because each vertex of K_n is connected by exactly one edge to each of the other $n-1$ vertices of K_n, the degree of each vertex of K_n is $n-1$. Thus the total degree of K_n equals the number of vertices times the degree of each vertex, or $n(n-1)$. Let e be the number of edges of K_n. By Theorem 10.1.1, the total degree of K_n equals $2e$, and so $n(n-1) = 2e$. Equivalently, $e = n(n-1)/2$ *[as was to be shown].*

36. *b.* $K_{1,3}$

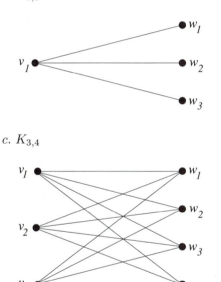

c. $K_{3,4}$

d. If $n \neq m$, the vertices of $K_{m,n}$ are divided into two groups: one of size m and the other of size n. Every vertex in the group of size m has degree n because each is connected to every vertex in the group of size n. So $K_{m,n}$ has n vertices of degree m. Similarly, every vertex in the group of size n has degree m because each is connected to every vertex in the group of size m. So $K_{m,n}$ has n vertices of degree m. Note that if $n = m$, then all $n + m = 2n$ vertices have the same degree, namely n.

e. The total degree of $K_{m,n}$ is $2mn$ because $K_{m,n}$ has m vertices of degree n (which contribute mn to its total degree) and n vertices of degree m (which contribute another mn to its total degree)

f. The number of edges of $K_{m,n} = mn$. The reason is that the total degree of $K_{m,n}$ is $2mn$, and so, by Theorem 10.1.1, $K_{m,n}$ has $2mn/2 = mn$ edges. Another way to reach this conclusion is to say that $K_{m,n}$ has n edges coming out of each of the group of m vertices (each leading to a vertex in the group of n vertices) for a total of mn edges. Equivalently, $K_{m,n}$ has m edges coming out of each of the group of n vertices (each leading to a vertex in the group of m vertices) for a total of mn edges.

39. *b.*

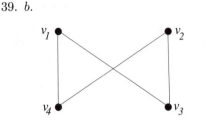

42. The graph obtained by taking all the vertices and edges of G together with all the edges of G' is K_n. Therefore, by exercise 33b, the number of edges of G plus the number of edges of G' equals $n(n-1)/2$.

45. Yes. Suppose that in a group of two or more people, each person is acquainted with a different number of people. Then the acquaintance graph representing the situation is a simple graph in which all the vertices have different degrees. But by exercise 44(c) such a graph does not exist. Hence the supposition is false, and so in a group of two or more people there must be at least two people who are acquainted with the same number of people within the group.

48. In the following graph each course number is represented as a vertex. Vertices are joined if, and only if, the corresponding courses have a student in common.

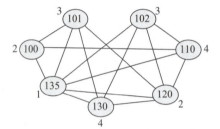

Vertex 135 has maximal degree, so use color #1 for it. All vertices share edges with vertex 135, and so color #1 cannot be used on any other vertex.

From the remaining uncolored vertices, only vertex 120 has maximal degree. So use color #2 for it. Because vertex 100 does not share an edge with vertex 120, color #2 may also be used for it.

From the remaining uncolored vertices, all of 101, 102, 110, and 130 have maximal degree. Choose any one of them, say vertex 101, and use color #3 for it. Neither vertex 102 nor vertex 110 shares an edge with vertex 101, but they do share an edge with each other. So color #3 may be used for only one of them. If color #3 is used for vertex 110, then, since the remaining vertices 130 and 102 are connected, two additional colors would be needed for them to have different colors. On the other hand, if color #3 is used for vertex 102, then, since the remaining vertices, 110 and 130, are not connected to each other, color 4 may be used for both. Therefore, to minimize the number of colors, color #3 should be used for vertex 102 and color #4 for vertices 110 and 130. The result is indicated in the annotations on the graph.

To use the results for scheduling exams, let color n correspond to exam time n. Then

Time 1: MCS135

Time 2: MCS 100 and MCS120

Time 3: MCS101 and MCS102

Time 4: MCS110 and MCS130

Note that because, for example, MSC135, MSC102, MSC110, and MSC 120 are all connected to each other, they must all be given different colors, and so the schedule for the seven exams must use at least four time periods.

Section 10.2

3. *b.* No, because $e_1 e_2$ could refer either to $v_1 e_1 v_2 e_2 v_1$ or to $v_2 e_1 v_1 e_2 v_2$.

6. *b.* $\{v_7, v_8\}, \{v_1, v_2\}, \{v_3, v_4\}$

   c. $\{v_2, v_3\}, \{v_6, v_7\}, \{v_7, v_8\}, \{v_9, v_{10}\}$

9. *b.* Yes, by Theorem 10.2.3 since G is connected and every vertex has even degree.

   c. Not necessarily. It is not specified that G is connected. For instance, the following graph satisfies the given conditions but does not have an Euler circuit:

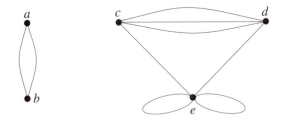

15. One Euler circuit is the following: *stuvwxyzrsuwyuzs*.

18. Yes. One Euler circuit is *ABDEACDA*.

21. One Euler path from u to w is $uv_1 v_2 v_3 uv_0 v_7 v_6 v_3 v_4 v_6 wv_5 v_4 w$.

24. One Hamiltonian circuit is *balkjedcfihgb*. The only other one traverses this circuit in the opposite direction.

27. Call the given graph G and suppose G has a Hamiltonian circuit. Then G has a subgraph H that satisfies conditions (1) – (4) of Proposition 10.2.6. Since the degree of B in G is five and every vertex in H has degree two, three edges incident on B must be removed from G to create H. Edge $\{B, C\}$ cannot be removed because doing so would result in vertex C having degree less than two in H. Similar reasoning shows that edges $\{B, E\}$, $\{B, F\}$, and $\{B, A\}$ cannot be removed either. It follows that the degree of B in H must be at least four, which contradicts the condition that every vertex in H has degree two in H. Hence no such subgraph H can exist, and so G does not have a Hamiltonian circuit.

30. One Hamiltonian circuit is $v_0 v_1 v_5 v_4 v_7 v_6 v_2 v_3 v_0$.

33. Other such graphs are those shown in exercises 17, 21, 23, 24, 29 and 30.

36. It is clear from the map that only a few routes have a chance of minimizing the distance. For instance, one must go to either Düsseldorf or Luxembourg just after leaving Brussels or just before returning to Brussels, and one must either travel from Berlin directly to Munich or the reverse. The possible minimizing routes are those shown below plus the same routes traveled in the reverse direction.

Route	Total Distance (in km)
Bru-Lux-Düss-Ber-Mun-Par-Bru	$219 + 224 + 564 + 585 + 832 + 308 = 2732$
Bru-Düss-Ber-Mun-Par-Lux-Bru	$223 + 564 + 585 + 832 + 375 + 219 = 2798$
Bru-Düss-Lux-Ber-Mun-Par-Bru	$223 + 224 + 764 + 585 + 832 + 308 = 2936$
Bru-Düss-Ber-Mun-Lux-Par-Bru	$223 + 564 + 585 + 517 + 375 + 308 = 2572$

The routes that minimize distance, therefore, are the bottom route shown in the table and that same route traveled in the reverse direction.

39. <u>Proof:</u>

Suppose vertices v and w are part of a circuit in a graph G and one edge e is removed from the circuit. Without loss of generality, we may assume the v occurs before the w in the circuit, and we may denote the circuit by $v_0 e_1 v_1 e_2 \ldots e_{n-1} v_{n-1} e_n v_0$ with $v_i = v$, $v_j = w$, $i < j$, and $e_k = e$.

In case either $k \le i$ or $k > j$, then $v = v_i e_{i+1} v_{i+1} \ldots v_{j-1} e_j v_j = w$ is a trail in G from v to w that does not include e.

In case $i < k \le j$, then $v = v_i e_i v_{i-1} e_{i-1} \ldots v_1 e_1 v_0 e_n v_{n-1} \ldots e_{j+1} v_j = w$ is a trail in G from v to w that does not include e.

These possibilities are illustrated by examples (1) and (2) in the diagram below. In both cases there is a trail in G from v to w that does not include e.

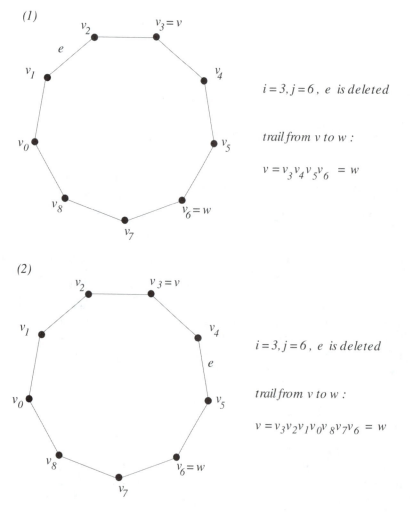

(1)

$i = 3, j = 6$, e is deleted

trail from v to w :

$v = v_3 v_4 v_5 v_6 = w$

(2)

$i = 3, j = 6$, e is deleted

trail from v to w :

$v = v_3 v_2 v_1 v_0 v_8 v_7 v_6 = w$

48. *a.* Let m and n be positive integers and let $K_{m,n}$ be a complete bipartite graph on (m, n) vertices. Since $K_{m,n}$ is connected, by Theorem 10.2.4 it has an Euler circuit if, and only if, every vertex has even degree. But $K_{m,n}$ has m vertices of degree n and n vertices of degree m. So $K_{m,n}$ has an Euler circuit if, and only if, both m and n are even.

b. Let m and n be positive integers, let $K_{m,n}$ be a complete bipartite graph on (m, n) vertices, and suppose $V_1 = \{v_1, v_2, \ldots, v_m\}$ and $V_2 = \{w_1, w_2, \ldots, w_n\}$ are the disjoint sets of vertices such that each vertex in V_1 is joined by an edge to each vertex in V_2 and no vertex within V_1 or V_2 is joined by an edge to any other vertex within the same set. If $m = n \ge 2$, then $K_{m,n}$ has the following Hamiltonian circuit: $v_1w_1v_2w_2\ldots v_mw_mv_1$. If $K_{m,n}$ has a Hamiltonian circuit, then $m = n$ because the vertices in any Hamiltonian circuit must alternate between V_1 and V_2 (since no edges connect vertices within either set) and because no vertex, except the first and last, appears twice in a Hamiltonian circuit. If $m = n = 1$, then $K_{m,n}$ does not have a Hamiltonian circuit because $K_{1,1}$ contains just one edge joining two vertices. Therefore, $K_{m,n}$ has a Hamiltonian circuit if, and only if, $m = n \geq 2$.

Section 10.3

3. *b.*

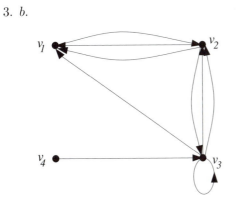

Any labels may be applied to the edges because the adjacency matrix does not determine edge labels.

6. *b.* The graph is not connected; the matrix shows that there are no edges joining the vertices from the set $\{v_1, v_2\}$ to those in the set $\{v_3, v_4\}$.

9. *b.*
$$\begin{bmatrix} 0 & 8 \\ -5 & 4 \end{bmatrix}$$
c.
$$\begin{bmatrix} -2 & -3 \\ 4 & 6 \end{bmatrix}$$

18. <u>Proof (by mathematical induction:</u> Let the property $P(n)$ be the sentence

$$\mathbf{A}^n \text{ is symmetric.} \qquad \leftarrow P(n)$$

We will prove that $P(n)$ is true for all integers $n \geq 1$.

Show that $P(1)$ is true: $P(1)$ is true because by assumption $\mathbf{A}$ is a symmetric matrix.

Show that for all integers $k \geq 1$, if $P(k)$ is true then $P(k+1)$ is true: Let k be any integer with $k \geq 1$, and suppose

$$\mathbf{A}^k \text{ is symmetric.} \qquad \leftarrow \begin{array}{c} P(k) \\ \text{inductive hypothesis} \end{array}$$

We must show that

$$\mathbf{A}^{k+1} \text{ is symmetric.} \qquad \leftarrow P(k+1)$$

Let $\mathbf{A}^k = (b_{ij})$. Then for all $i, j = 1, 2, \ldots, m$,

$$
\begin{array}{lll}
\text{the } ij\text{th entry of } \mathbf{A}^{k+1} & = & \text{the } ij\text{th entry of } \mathbf{A}\mathbf{A}^k \qquad \text{by definition of matrix power} \\[4pt]
& = & \displaystyle\sum_{r=1}^{m} a_{ir}b_{rj} \qquad\qquad \text{by definition of matrix multiplication} \\[6pt]
& = & \displaystyle\sum_{r=1}^{m} a_{ri}b_{jr} \qquad\qquad \begin{array}{l}\text{because } A \text{ is symmetric by hypothesis and} \\ A^k \text{is symmetric by inductive hypothesis}\end{array} \\[6pt]
& = & \displaystyle\sum_{r=1}^{m} b_{jr}a_{ri} \qquad\qquad \begin{array}{l}\text{because multiplication of real numbers} \\ \text{is commutative}\end{array} \\[6pt]
& = & \text{the } ji\text{th entry of } \mathbf{A}^k\mathbf{A} \quad\ \text{by definition of matrix multiplication} \\[2pt]
& = & \text{the } ji\text{th entry of } \mathbf{A}\mathbf{A}^k \quad\ \text{by exercise 17} \\[2pt]
& = & \text{the } ji\text{th entry of } \mathbf{A}^{k+1} \ \text{by definition of matrix power.}
\end{array}
$$

Therefore, $\mathbf{A}^{k+1}$ is symmetric *[as was to be shown]*.

21. <u>Proof (by mathematical induction:</u> Let the property $P(n)$ be the sentence

> All the entries along the main diagonal of $\mathbf{A}^n$ are equal to each other and all the entries off the main diagonal are also equal to each other. $\qquad \leftarrow P(n)$

We will prove that $P(n)$ is true for all integers $n \geq 1$.

Show that $P(1)$ is true: $P(1)$ is true because

$$
\mathbf{A}^1 = \mathbf{A} = \begin{bmatrix} 0 & 1 & 1 \\ 1 & 0 & 1 \\ 1 & 1 & 0 \end{bmatrix},
$$

which is the adjacency matrix for K_3, and all the entries along the main diagonal of $\mathbf{A}$ are 0 *[because K_3 has no loops]* and all the entries off the main diagonal are 1 *[because each pair of vertices is connected by exactly one edge]*.

Show that for all integers $m \geq 1$, if $P(m)$ is true then $P(m+1)$ is true: Let m be any integer with $m \geq 1$, and suppose

> All the entries along the main diagonal of $\mathbf{A}^m$ are equal to each other and all the entries off the main diagonal are also equal to each other.
> $\qquad \begin{array}{l}P(m) \\ \leftarrow \text{ inductive} \\ \text{hypothesis}\end{array}$

We must show that

> All the entries along the main diagonal of $\mathbf{A}^{m+1}$ are equal to each other and all the entries off the main diagonal are also equal to each other. $\qquad \leftarrow P(m+1)$

By inductive hypothesis,

$$
\mathbf{A}^m = \begin{bmatrix} b & c & c \\ c & b & c \\ c & c & b \end{bmatrix} \text{ for some integers } b \text{ and } c.
$$

It follows that

$$
\mathbf{A}^{m+1} = \mathbf{A}\mathbf{A}^m = \begin{bmatrix} 0 & 1 & 1 \\ 1 & 0 & 1 \\ 1 & 1 & 0 \end{bmatrix}\begin{bmatrix} b & c & c \\ c & b & c \\ c & c & b \end{bmatrix} = \begin{bmatrix} 2c & b+c & b+c \\ b+c & 2c & b+c \\ b+c & b+c & 2c \end{bmatrix}
$$

As can be seen, all the entries of $\mathbf{A}^{m+1}$ along the main diagonal are equal to each other and all the entries off the main diagonal are equal to each other. So the property is true for $n = m+1$.

Section 10.4

3. The graphs are isomorphic. One way to define to isomorphism is as follows.

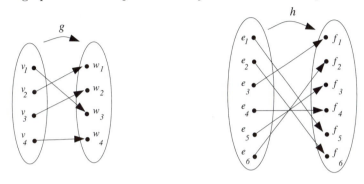

9. The graphs are isomorphic. One way to define to isomorphism is as follows.

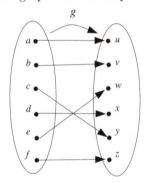

15. Of all nonisomorphic simple graphs with four vertices, there is one with 0 edges, one with 1 edge, two with 2 edges, three with 3 edges, two with 4 edges, one with 5 edges, and one with 6 edges. These eleven graphs are shown below.

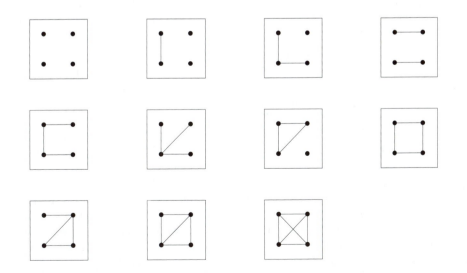

18. There are three nonisomorphic graphs with four vertices and three edges in which all 3 edges are loops, five in which 2 edges are loops and 1 is not a loop, six in which 1 edge is a loop and

2 edges are not loops, and six in which none of the 3 edges is a loop. These twenty graphs are shown below.

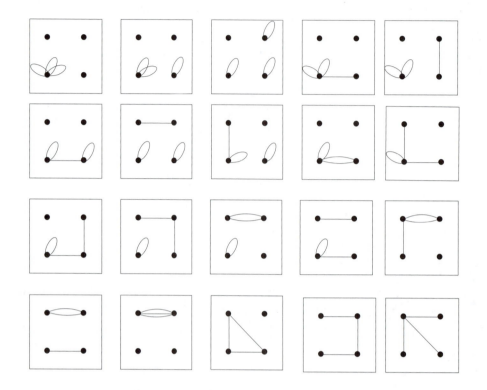

24. Proof:

Suppose G and G' are isomorphic graphs and suppose G has a simple circuit C of length k, where k is a nonnegative integer. By definition of graph isomorphism, there are one-to-one correspondences $g: V(G) \to V(G')$ and $h: E(G) \to E(G')$ that preserve the edge-endpoint functions in the sense that for all v in $V(G)$ and e in $E(G)$, v is an endpoint of $e \Leftrightarrow g(v)$ is an endpoint of $h(e)$.

Let C be $v_0 e_1 v_1 e_2 \ldots e_k v_k(= v_0)$, and let C' be $g(v_0)h(e_1)g(v_1)h(e_2)\ldots h(e_k)g(v_k)(= g(v_0))$. By the same reasoning as in the solution to exercise 23 in Appendix B, C' is a circuit of length k in G'.

Suppose C' is not a simple circuit. Then C' has a repeated vertex, say $g(v_i) = g(v_j)$ for some $i, j = 0, 1, 2, \ldots, k-1$ with $i \neq j$. But since g is a one-to-one correspondence this implies that $v_i = v_j$, which is impossible because C is a simple circuit. Hence the supposition is false, and so we conclude that C' is a simple circuit. Therefore G' has a simple circuit of length k.

27. Proof:

Suppose G and G' are isomorphic graphs and suppose G is connected. By definition of graph isomorphism, there are one-to-one correspondences $g: V(G) \to V(G')$ and $h: E(G) \to E(G')$ that preserve the edge-endpoint functions in the sense that for all v in $V(G)$ and e in $E(G)$, v is an endpoint of $e \Leftrightarrow g(v)$ is an endpoint of $h(e)$. Suppose w and x are any two vertices of G'. Then $u = g^{-1}(w)$ and $v = g^{-1}(x)$ are distinct vertices in G (because g is a one-to-one correspondence). Since G is connected, there is a walk in G connecting u and v. Say this walk is $u e_1 v_1 e_2 v_2 \ldots e_n v$. Because g and h preserve the edge-endpoint functions, $w = g(u)h(e_1)g(v_1)h(e_2)g(v_2)\ldots h(e_n)g(v) = x$ is a walk in G' connecting w and x.

30. Suppose that G and G' are isomorphic via one-to-one correspondences $g: V(G) \to V(G')$ and $h: E(G) \to E(G')$, where g and h preserve the edge-endpoint functions. Now w_6 has degree

one in G', and so by the argument given in Example 10.4.4, w_6 must correspond to one of the vertices of degree one in G: either $g(v_1) = w_6$ or $g(v_6) = w_6$. Similarly, since w_5 has degree three in G', w_5 must correspond to one of the vertices of degree three in G: either $g(v_3) = w_5$ or $g(v_4) = w_5$. Because g and h preserve the edge-endpoint functions, edge f_6 with endpoints w_5 and w_6 must correspond to an edge in G with endpoints v_1 and v_3, or v_1 and v_4, or v_6 and v_3, or v_6 and v_4. But this contradicts the fact that none of these pairs of vertices are connected by edges in G. Hence the supposition is false, and G and G' are not isomorphic.

Section 10.5

3. By Theorem 10.5.2, a tree with n vertices (where $n \geq 1$) has $n - 1$ edges, and so by Theorem 10.1.1, its total degree is twice the number of edges, or $2(n - 1) = 2n - 2$.

6. Define an infinite graph G as follows: $V(G) = \{v_i \mid i \in \mathbf{Z}\} = \{\ldots, v_{-2}, v_{-1}, v_0, v_1, v_2, \ldots\}$, $E(G) = \{e_i \mid i \in \mathbf{Z}\} = \{\ldots, e_{-2}, e_{-1}, e_0, e_1, e_2, \ldots\}$, and the edge-endpoint function is defined by the rule $f(e_i) = \{v_{i-1}, v_i\}$ for all $i \in \mathbf{Z}$. Then G is circuit-free, but each vertex has degree two. G is illustrated below.

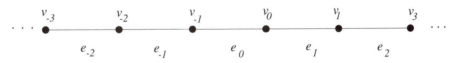

15. One circuit-free graph with seven vertices and four edges is shown below.

18. Any tree with five vertices has four edges. By Theorem 10.1.1, the total degree of such a graph is eight, not ten. Hence there is no tree with five vertices and total degree ten.

21. Any tree with ten vertices has nine edges. By Theorem 10.1.1, the total degree of such a tree is 18, not 24. Hence there is no such graph.

24. Yes. Given any two vertices u and w of G', then u and w are vertices of G neither equal to v. Since G is connected, there is a walk in G from u to w, and so by Lemma 10.2.1, there is a path in G from u to w. This path does not include edge e or vertex v because a path does not have a repeated edge, and e is the unique edge incident on v. *[If a path from u to w leads into v, then it must do so via e. But then it cannot emerge from v to continue on to w because no edge other than e is incident on v.]* Thus this path is a path in G'. It follows that any two vertices of G' are connected by a walk in G', and so G' is connected.

30. A tree with five vertices must have four edges and, therefore, a total degree of 8. Since at least two vertices have degree 1 and no vertex has degree greater than 4, the possible degrees of the five vertices are as follows: 1,1,1,1,4; 1,1,1,2,3; and 1,1,2,2,2. The corresponding trees are shown below.

Section 10.6

3. *b.*

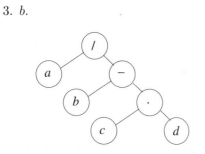

12. There is no tree with the given properties because any full binary tree with eight internal vertices has nine terminal vertices, not seven.

15. There is no tree with the given properties because a full binary tree with five internal vertices has $2 \cdot 5 + 1$ or eleven vertices in all, not nine.

18. There is no full binary tree with sixteen vertices because a full binary tree has $2k + 1$ vertices, where k is the number of internal vertices, and $16 \neq 2k + 1$ for any integer k.

Section 10.7

6. Minimum spanning tree:

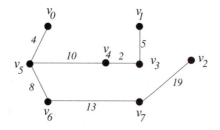

Order of adding the edges: $\{v_3, v_4\}$, $\{v_0, v_5\}$, $\{v_1, v_3\}$, $\{v_5, v_6\}$, $\{v_4, v_5\}$, $\{v_6, v_7\}$, $\{v_2, v_7\}$

15.

Step	newpage(T)	$E(T)$	F	$L(a)$	$L(b)$	$L(c)$	$L(d)$	$L(e)$	$L(f)$	$L(g)$
0	$\{a\}$	$\emptyset$	$\{a\}$	0	∞	∞	∞	∞	∞	∞
1	$\{a\}$	$\emptyset$	$\{b, e, g\}$	0	**3***	∞	∞	3	∞	4
2	$\{a, b\}$	$\{\{a, b\}\}$	$\{c, e, g\}$	0	3	10	∞	**3**	∞	4
3	$\{a, b, e\}$	$\{\{a, b\}, \{a, e\}\}$	$\{c, d, f, g\}$	0	3	10	14	3	7	**4**
4	$\{a, b, e, g\}$	$\{\{a, b\}, \{a, e\}, \{a, g\}\}$	$\{c, d, f\}$	0	3	10	14	3	**5**	4
5	$\{a, b, e, g, f\}$	$\{\{a, b\}, \{a, e\}, \{a, g\}, \{g, f\}\}$								

*At this point, vertex e could have been chosen instead of vertex b.

18. *a.* If there were two distinct paths from one vertex of a tree to another, they (or pieces of them) could be patched together to obtain a circuit. But a tree cannot have a circuit.

21. *b.* <u>Counterexample:</u> Let G be the following simple graph.

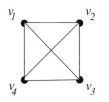

Then G has the spanning trees shown below.

These trees have no edge in common.

27. <u>Proof:</u> Suppose not. That is, suppose there exists a connected, weighted graph G with n vertices and an edge e that (1) has larger weight than any other edge of G, (2) is in a circuit C of G, and (3) is in a minimum spanning tree T for G. Let the endpoints of e be vertices v and w, and let H be the graph obtained from T by removing e. In other words, $V(H) = V(T)$ and $E(H) = E(T) - \{e\}$. Then H is a circuit-free subgraph of T that contains all the vertices of G but only $n - 2$ edges, too few to be a tree. Now H consists of two components, one, say H_v, containing v and the other, say H_w, containing w. Let e' be a "bridge" from H_w to H_v. That is, as shown in the solution to exercise 39 in Section 10.2, there is a trail in G from v to w that does not include e, and so we may let e' be the edge in the trail that immediately precedes the first vertex in the trail that is in H_v. Let T' be the graph obtained from H by adding e'. More precisely, $V(T') = V(H)$ and $E(T') = E(H) \cup \{e\}$. Then T' is connected, contains every vertex of G (as does T), and has $n - 1$ edges (the same as T). Hence, by Theorem 10.5.4, T' is a spanning tree for G. Now

$$w(T') = w(T) - w(e) + w(e') = w(T) - (w(e) - w(e')) < w(T)$$

because $w(e) > w(e')$. Thus T' is a spanning tree of smaller weight than a minimum spanning tree for G, which is a contradiction. Hence the supposition is false, and the given statement is true.

30. <u>Proof:</u> Suppose that G is a connected, weighted graph with n vertices and that T is the output graph produced when G is input to Algorithm 10.7.4. Clearly T is a subgraph of G and T is connected because no edge is removed from T as T is being constructed if its removal would disconnect T. Also T is circuit-free because if T had a circuit then the circuit would contain edges $e_1, e_2, \ldots, e_k$ of maximal weight. At some point during execution of the algorithm, each of these edges would be examined (since all edges are examined eventually). Let e_i be the first such edge to be examined. When examined, e_i must be removed because deletion of an edge from a circuit does not disconnect a graph and at the time e_i is examined no other edge of the circuit would have been removed. But this contradicts the supposition that e_i was one of the edges in the output graph T. Thus T is circuit-free. Furthermore, T contains every vertex of G since only edges, not vertices, are removed from G in the construction of T. Hence T is a spanning tree for G.

Next we show that T has minimum weight. Let T_1 be any minimum spanning tree for G such that the number of edges T_1 and T have in common is a maximum. If $T = T_1$, we are done. So suppose $T \neq T_1$. Then there is an edge e of T that is not in T_1. *[Since trees T and*

T_1 both have the same vertex set, if they differ at all, they must have different, but same-size, edge sets.] Now adding e to T_1 produces a graph with a unique circuit (exercise 19). Let e' be an edge of this circuit such that e' is not in T. *[Such an edge must exist because T is a tree and hence circuit-free.]* Let T_2 be the graph obtained from T_1 by removing e' and adding e. Note that T_2 has n vertices and $n-1$ edges and that T_2 is connected *[since, by Lemma 10.5.3, the subgraph obtained by removing an edge from a circuit in a connected graph is connected].* Consequently, T_2 is a spanning tree for G. In addition,

$$w(T_2) = w(T_1) - w(e') + w(e).$$

Now $w(e) \leq w(e')$ because at the stage in Algorithm 10.7.4 when e' was removed, e could have been removed, and it would have been removed if $w(e) > w(e')$. Thus

$$w(T_2) \;=\; w(T_1) - \underbrace{w(e') + w(e)}_{\geq\, 0} \;\leq\; w(T_1).$$

But T_1 is minimum spanning tree for G, and thus, since T_2 is a spanning tree with weight less than or equal to the weight of T_1, T_2 is also a minimum spanning tree for G.

Finally note that by construction, T_2 has one more edge in common with T than T_1 does, which contradicts the choice of T_1 as a minimum spanning tree for G with a maximum number of edges in common with T. Thus the supposition that $T \neq T_1$ is false, and hence T itself is a minimum spanning tree for G.

Review Guide: Chapter 10

Definitions: How are the following terms defined?

- graph, edge-endpoint function *(p. 626)*
- loop in a graph, parallel edges, adjacent edges, isolated vertex, edge incident on an endpoint *(p. 626)*
- directed graph *(p. 629)*
- simple graph *(p. 632)*
- complete graph on n vertices *(p. 633)*
- complete bipartite graph on (m, n) vertices *(p. 633)*
- subgraph *(p. 634)*
- degree of a vertex in a graph, total degree of a graph *(p. 635)*
- walk, trail, path, closed walk, circuit, simple circuit *(p. 644)*
- connected vertices, connected graph *(p. 646)*
- connected component of a graph *(p. 647)*
- Euler circuit in a graph *(p. 648)*
- Euler trail in a graph *(p. 652)*
- Hamiltonian circuit in a graph *(p. 654)*
- adjacency matrix of a directed (or undirected) graph *(p. 662)*
- symmetric matrix *(p. 664)*
- $n \times n$ identity matrix *(p. 669)*
- powers of a matrix *(p. 670)*
- isomorphic graphs *(p. 676)*
- isomorphic invariant for graphs *(p. 679)*
- circuit-free graph *(p. 683)*
- tree, forest, trivial tree *(p. 683)*
- parse tree, syntactic derivation tree *(p. 684)*
- terminal vertex (or leaf), internal vertex (or branch vertex) *(p. 688)*
- rooted tree, level of a vertex in a rooted tree, height of a rooted tree *(p. 694)*
- parents, children, siblings, descendants, and ancestors in a rooted tree *(p. 694)*
- binary tree, full binary tree, subtree *(p. 696)*
- spanning tree *(p. 702)*
- weighted graph, minimum spanning tree *(p. 704)*

Graphs

- How can you use a graph as a model to help solve a problem? *(p. 631)*
- What does the handshake theorem say? In other words, how is the total degree of a graph related to the number of edges of the graph? *(p. 636)*
- How can you use the handshake theorem to determine whether graphs with specified properties exist? *(pp. 636-638)*
- If an edge is removed from a circuit in a graph, does the graph remain connected? *(p. 647, 690)*
- A graph has an Euler circuit if, and only if, it satisfies what two conditions? *(p. 652)*
- A graph has a Hamiltonian circuit if, and only if, it satisfies what four conditions? *(p. 655)*
- What is the traveling salesman problem? *(p. 656)*
- How do you find the adjacency matrix of a directed (or undirected) graph? How do you find the graph that corresponds to a given adjacency matrix? *(p. 663)*

- How can you determine the connected components of a graph by examining the adjacency matrix of the graph? *(p. 666)*
- How do you multiply two matrices? *(p. 666)*
- How do you use matrix multiplication to compute the number of walks from one vertex to another in a graph? *(p. 672)*
- How do you show that two graphs are isomorphic? *(p. 677)*
- What are some invariants for graph isomorphisms? *(p. 679)*
- How do you establish that two simple graphs are isomorphic? *(p. 680)*

Trees

- How do you show that a saturated carbon molecule with k carbon atoms has $2k + 2$ hydrogen atoms? *(p. 686 and exercise 4 in Section 10.5)*
- If a tree has more than one vertex, how many vertices of degree 1 does it have? Why? *(p. 687)*
- If a tree has n vertices, how many edges does it have? Why? *(p. 688)*
- If a connected graph has n vertices, what additional property guarantees that it will be a tree? Why? *(p. 692)*
- How can you represent an algebraic expression using a binary tree? *(p. 696)*
- Given a full binary tree, what is the relation among the number of its internal vertices, terminal vertices, and total number of vertices? *(p. 697)*
- Given a binary tree, what is the relation between the number of its terminal vertices and its height? *(p. 698)*
- What is the relation between the number of edges in two different spanning trees for a graph? *(p. 702)*
- How does Kruskal's algorithm work? *(p. 704)*
- How do you know that Kruskal's algorithm produces a minimum spanning tree? *(p. 706)*
- How does Prim's algorithm work? *(p. 707)*
- How do you know that Prim's algorithm produces a minimum spanning tree? *(p. 708)*
- How does Dijkstra's shortest path algorithm work? *(p. 711)*
- How do you know that Dijkstra's shortest path algorithm produces a shortest path? *(p. 713)*

Chapter 11: Analysis of Algorithm Efficiency

The focus of Chapter 11 is the analysis of algorithm efficiency in Sections 11.3 and 11.5. The chapter opens with a brief review of the properties of function graphs that are especially important for understanding O-, Ω-, and Θ-notations, which are introduced in Section 11.2. For simplicity, the examples in Section 11.2 are restricted to polynomial and rational functions. Section 11.3 introduces the analysis of algorithm efficiency with examples that include sequential search, insertion sort, selection sort (in the exercises), and polynomial evaluation (in the exercises). Section 11.4 discusses the properties of logarithms that are particularly important in the analysis of algorithms and other areas of computer science, and Section 11.5 applies the properties to analyze algorithms whose orders involve logarithmic functions. Examples in Section 11.5 include binary search and merge sort.

Section 11.1

9.

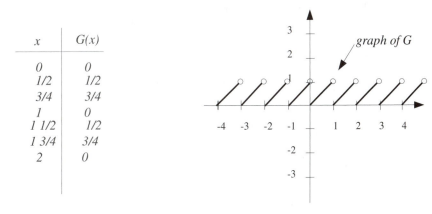

x	$G(x)$
0	0
1/2	1/2
3/4	3/4
1	0
1 1/2	1/2
1 3/4	3/4
2	0

18. *b.* When $x < 0$, k is increasing.

   Proof:

   Suppose $x_1 < x_2 < 0$. Multiplying both sides of this inequality by -1 gives $-x_1 > -x_2$, and adding $x_1 x_2$ to both sides gives $x_1 x_2 - x_1 > x_1 x_2 - x_2$. Now, since x_1 and x_2 are both negative, $x_1 x_2$ is positive, and hence

   $$\frac{x_1 x_2 - x_1}{x_1 x_2} > \frac{x_1 x_2 - x_2}{x_1 x_2}.$$

   Simplifying the two fractions gives

   $$\frac{x_2 - 1}{x_2} > \frac{x_1 - 1}{x_1}$$

   and so $k(x_1) < k(x_2)$ by definition of k.

21. *b.* Proof by contradiction:

   Suppose that g is not increasing. Then there exist real numbers x_1 and x_2 such that $0 < x_1 < x_2$ and $g(x_1) \geq g(x_2)$. By definition of g,

   $$x_1^{\frac{m}{n}} \geq x_2^{\frac{m}{n}}.$$

   Applying part (a) to this inequality gives

   $$\left(x_1^{\frac{m}{n}}\right)^n \geq \left(x_2^{\frac{m}{n}}\right)^n.$$

169

By the laws of exponents, $x_1^{\frac{m}{n} \cdot n} = x_1{}^m$ and $x_2^{\frac{m}{n} \cdot n} = x_2{}^m$, and so

$$x_1{}^m \geq x_2{}^m.$$

But, by part (a), $x_1{}^m < x_2{}^m$, and so we have reached a contradiction. Hence the supposition is false, and thus g is increasing.

Section 11.2

3. *a. Formal version of negation*: $f(x)$ is not $\Theta(g(x))$ if, and only if, $\forall$ positive real numbers k, A, and B, $\exists$ a real number $x > k$ such that either $|f(x)| < A\,|g(x)|$ or $|f(x)| > B\,|g(x)|$.

b. Informal version of negation: $f(x)$ is not $O(g(x))$ if, and only if, no matter what positive real numbers k, A, and B might be chosen, it is possible to find a real number x greater than k with the property that either $|f(x)| < A\,|g(x)|$ or $|f(x)| > B\,|g(x)|$.

9. Let $A = 1/2$, $B = 3$, and $k = 33$. Then by substitution, $A|x^2| \leq |3x^2 - 80x + 7| \leq B|x^2|$ for all $x > k$, and hence by definition of Θ-notation, $3x^2 - 80x + 7$ is $\Theta(x^2)$.

15. *a.* <u>Proof (by mathematical induction)</u>: Let the property $P(n)$ be the sentence

> If x is any real number with $x > 1$, then $x^n > 1$. $\leftarrow P(n)$

Show that $P(1)$ is true: We must show that if x is any real number with $x > 1$, then $x^1 > 1$. But this is true because $x^1 = x$. So $P(1)$ is true.

Show that for all integers $k \geq 1$, if $P(k)$ is true then $P(k+1)$ is true: Let k be any integer with $k \geq 1$, and suppose that

> If x is any real number with $x > 1$, then $x^k > 1$. $\leftarrow$ $\begin{array}{l} P(k) \\ \text{inductive hypothesis} \end{array}$

We must show that

> If x is any real number with $x > 1$, then $x^{k+1} > 1$. $\leftarrow P(k+1)$

So suppose x is any real number with $x > 1$. By inductive hypothesis, $x^k > 1$, and multiplying both sides by the positive number x gives $x \cdot x^k > x \cdot 1$, or, equivalently, $x^{k+1} > x$. But $x > 1$, and so, by transitivity of order, $x^{k+1} > 1$ *[as was to be shown]*.

b. <u>Proof</u>:

Suppose x is any real number with $x > 1$ and m and n are integers with $m < n$. Then $n - m$ is an integer with $n - m \geq 1$, and so, by part (a), $x^{n-m} > 1$. Multiplying both sides by x^m gives $x^m \cdot x^{n-m} > x^m \cdot 1$, and so, by the laws of exponents, $x^n > x^m$ *[as was to be shown]*.

21. *a.* For any real number $x > 1$,

$$\begin{aligned} |\lfloor \sqrt{x} \rfloor| \;&=\; \lfloor \sqrt{x} \rfloor &&\text{because since } x > 1 > 0, \text{ then } \lfloor \sqrt{x} \rfloor > 0 \\ &\leq\; \sqrt{x} &&\text{because } \lfloor r \rfloor \leq r \text{ for all real numbers } r \\ &=\; |\sqrt{x}| &&\text{because } \sqrt{x} \geq 0. \end{aligned}$$

Therefore, by transitivity of equality and order, $|\lfloor \sqrt{x} \rfloor| \leq |\sqrt{x}|$.

b. Suppose x is any real number with $x > 1$. By definition of floor,

$$\lfloor \sqrt{x} \rfloor \leq \sqrt{x} < \lfloor \sqrt{x} \rfloor + 1.$$

Now

$$\lfloor\sqrt{x}\rfloor + 1 \;\le\; 2\lfloor\sqrt{x}\rfloor$$

$\Leftrightarrow\; 1 \;\le\; \lfloor\sqrt{x}\rfloor \quad$ by subtracting $\lfloor\sqrt{x}\rfloor$ from both sides

$\Leftrightarrow\; 1 \;\le\; \sqrt{x} \quad$ by definition of floor

$\Leftrightarrow\; 1 \;\le\; x \quad$ by squaring both sides (okay because x is positive),

and the last inequality is true because we are assuming that $x > 1$. Thus,

$$\sqrt{x} < \lfloor\sqrt{x}\rfloor + 1 \quad\text{and}\quad \lfloor\sqrt{x}\rfloor + 1 \le 2\lfloor\sqrt{x}\rfloor,$$

and so, by the transitivity of order (Appendix A, T18),

$$\sqrt{x} \le 2\lfloor\sqrt{x}\rfloor.$$

Dividing both sides by 2 gives

$$\frac{1}{2}\sqrt{x} \le \lfloor\sqrt{x}\rfloor.$$

Finally, because all quantities are positive, we conclude that

$$\frac{1}{2}\left|\sqrt{x}\right| \le \left|\lfloor\sqrt{x}\rfloor\right|.$$

c. Let $A = \frac{1}{2}$ and $a = 1$. Then by substitution,

$$A\left|\sqrt{x}\right| \le \left|\lfloor\sqrt{x}\rfloor\right| \quad\text{for all } x > a,$$

and hence by definition of Ω-notation, $\lfloor\sqrt{x}\rfloor$ is $\Omega(\sqrt{x})$.

Let $B = 1$ and $b = 1$. Then

$$\left|\lfloor\sqrt{x}\rfloor\right| \le B\left|\sqrt{x}\right|$$

for all real numbers $x > b$, and so by definition of O-notation, $\lfloor\sqrt{x}\rfloor$ is $O(\sqrt{x})$.

d. By part (c) and Theorem 11.2.1(1), we can immediately conclude that $\lfloor\sqrt{x}\rfloor$ is $\Theta(\sqrt{x})$.

24. *a.* For all real numbers $x > 1$,

$$\left|\tfrac{1}{4}x^5 - 50x^3 + 3x + 12\right| \;\le\; \left|\tfrac{1}{4}x^5\right| + \left|50x^3\right| + \left|3x\right| + \left|12\right| \qquad \text{by the triangle inequality}$$

$$= \tfrac{1}{4}x^5 + 50x^3 + 3x + 12 \qquad \text{because } \tfrac{1}{4}x^5,\ 50^3,\ 3x,\\ \text{and } 12 \text{ are positive}$$

$$\le \tfrac{1}{4}x^5 + 50x^5 + 3x^5 + 12x^5 \qquad \text{because } x^3 < x^5,\ x < x^5,\\ \text{and } 1 < x^5 \text{ for } x > 1$$

$$\le 66x^5 \qquad \text{because } \tfrac{1}{4} + 50 + 3 + 12 < 66$$

$$= 66|x^5| \qquad \text{because } x^5 \text{ is positive.}$$

Therefore, by transitivity of equality and order, $\left|\tfrac{1}{4}x^5 - 50x^3 + 3x + 12\right| \le 66|x^5|$.

b. Let $B = 66$ and $b = 1$. Then by substitution, $\left|\tfrac{1}{4}x^5 - 50x^3 + 3x + 12\right| \le B|x^2|$ for all $x > b$. Hence by definition of O-notation,

$$\frac{1}{4}x^5 - 50x^3 + 3x + 12 \text{ is } O(x^2).$$

27. <u>Proof</u>:

Suppose $a_0, a_1, a_2, \ldots, a_n$ are real numbers and $a_n \neq 0$; and let

$$d = 2\left(\frac{|a_0| + |a_1| + |a_2| + \cdots + |a_{n-1}|}{|a_n|}\right).$$

Let a be greater than or equal to the maximum of d and 1. Then if $x > a$

$$x \geq 2\left(\frac{|a_0| + |a_1| + |a_2| + \cdots + |a_3| + |a_{n-1}|}{|a_n|}\right)$$

$$\Rightarrow \qquad \frac{1}{2}|a_n|x \geq |a_0| + |a_1| + |a_2| + \cdots + |a_{n-1}|$$

by multiplying both sides by $\frac{1}{2}|a_n|$

$$\Rightarrow \qquad (1 - \frac{1}{2})|a_n|x \geq |a_0| \cdot \frac{1}{x^{n-1}} + |a_1| \cdot \frac{1}{x^{n-2}} + |a_2| \cdot \frac{1}{x^{n-3}} + \cdots + |a_{n-2}| \cdot \frac{1}{x} + |a_{n-1}| \cdot 1$$

because by exercise 15, when $x > 1$ and $m \geq 1$,
then $x^m > 1$, and so $1 > \frac{1}{x^m}$

$$\Rightarrow \qquad |a_n|x^n - \frac{|a_n|}{2}x^n \geq |a_0| + |a_1|x + |a_2|x^2 + \cdots + |a_{n-2}|x^{n-2} + |a_{n-1}|x^{n-1}$$

by multiplying both sides by x^{n-1}.

Subtracting all terms on the right-hand side from both sides and adding the second term on the left-hand side to both sides gives

$$|a_n|x^n - |a_{n-1}|x^{n-1} - |a_{n-2}|x^{n-2} - \cdots - |a_2|x^2 - |a_1|x - |a_0| \geq \frac{|a_n|}{2}x^n. \qquad (*)$$

Now, by the triangle inequality, for all real numbers r and s,

$$|r| = |(r+s) + (-s)| \leq |r+s| + |-s| = |r+s| + |s|,$$

and thus,

$$|r| - |s| \leq |r+s|.$$

It follows by repeated application of this result that, when $x > 1$,

$$|a_n|x^n - |a_{n-1}|x^{n-1} - |a_{n-2}|x^{n-2} - \cdots - |a_2|x^2 - |a_1|x - |a_0|$$
$$= |a_nx^n| - |a_{n-1}x^{n-1}| - |a_{n-2}x^{n-2}| - \cdots - |a_2x^2| - |a_1x| - |a_0|$$
$$\leq |a_nx^n + a_{n-1}x^{n-1} + a_{n-2}x^{n-2} + \cdots + a_2x^2 + a_1x + a_0|. \qquad (**)$$

Using the transitive property of order to combine $(*)$ and $(**)$ gives that

$$\frac{|a_n|}{2}x^n \leq |a_nx^n + a_{n-1}x^{n-1} + a_{n-2}x^{n-2} + \cdots + a_2x^2 + a_1x + a_0|.$$

Let $A = \frac{|a_n|}{2}$ and let a be as defined above. Then

$$Ax^n \leq |a_nx^n + a_{n-1}x^{n-1} + a_{n-2}x^{n-2} + \cdots + a_2x^2 + a_1x + a_0| \qquad \text{for all real numbers } x > a.$$

It follows by definition of Ω-notation that

$$a_nx^n + a_{n-1}x^{n-1} + a_{n-2}x^{n-2} + \cdots + a_2x^2 + a_1x + a_0 \text{ is } \Omega(x^n).$$

30. Let $a = 2\left(\dfrac{50 + 3 + 12}{1/4}\right) = 520$, and let $A = \dfrac{1}{2} \cdot \dfrac{1}{4} = \dfrac{1}{8}$. If $x > 520$, then

$$x \geq 2\left(\frac{50 + 3 + 12}{1/4}\right)$$

$$\Rightarrow \qquad \frac{1}{2} \cdot \frac{1}{4} x \geq 50 + 3 + 12$$

by multiplying both sides by $\frac{1}{2} \cdot \frac{1}{4}$

$$\Rightarrow \qquad (1 - \frac{1}{2}) \cdot \frac{1}{4} x \geq 50 \frac{1}{x} + 3 \cdot \frac{1}{x^3} + 12 \cdot \frac{1}{x^4}$$

because $1 - \frac{1}{2} = \frac{1}{2}$, and, since $x > 520 > 1$,
then $1 > \frac{1}{x}$, $1 > \frac{1}{x^3}$ and $1 > \frac{1}{x^4}$

$$\Rightarrow \qquad \frac{1}{4} x^5 - \frac{1}{2} \frac{1}{4} x^5 \geq 50 x^3 + 3x + 12$$

by multiplying both sides by x^4

$$\Rightarrow \qquad \frac{1}{4} x^5 - 50 x^3 - 3x - 12 \geq \frac{1}{2} \cdot \frac{1}{4} x^5$$

by subtracting $50 x^3 + 3x + 12$ from and
adding $\frac{1}{2} \cdot \frac{1}{4} x^5$ to both sides

$$\Rightarrow \qquad \frac{1}{4} x^5 - 50 x^3 + 3x + 12 \geq \frac{1}{2} \cdot \frac{1}{4} x^5$$

because $3x + 12 > -3x - 12$ since $x > 0$

$$\Rightarrow \qquad \left| \frac{1}{4} x^5 - 50 x^3 + 3x + 12 \right| \geq \frac{1}{8} \left| x^5 \right|$$

because both sides are nonnegative.

Thus for all real numbers $x > a$, $\left| \dfrac{1}{4} x^5 - 50 x^3 + 3x + 12 \right| \geq A \left| x^5 \right|$. Hence, by definition of Ω-notation, we conclude that $\dfrac{1}{4} x^5 - 50 x^3 + 3x + 12$ is $\Omega(x^5)$.

33. By exercise 24, $\dfrac{1}{4} x^5 - 50 x^3 + 3x + 12$ is $O(x^5)$ and, by exercise 31, $\dfrac{1}{4} x^5 - 50 x^3 + 3x + 12$ is $\Omega(x^5)$. Thus, by Theorem 11.2.1(1), $\dfrac{1}{4} x^5 - 50 x^3 + 3x + 12$ is $\Theta(x^5)$.

36. Note that

$$\frac{x(x-1)}{2} + 3x = \frac{x^2 - x}{2} + \frac{6}{2} x$$

$$= \frac{1}{2} x^2 + \frac{5}{2} x \qquad \text{by algebra,}$$

and so, by the theorem on polynomial orders, $\dfrac{x(x-1)}{2} + 3x$ is $\Theta(x^2)$

39. Note that

$$2(n-1) + \frac{n(n+1)}{2} + 4\left(\frac{n(n-1)}{2}\right) = 2n - 2 + \frac{n^2}{2} + \frac{n}{2} + 2(n^2 - n)$$

$$= \frac{5}{2} n^2 + \frac{1}{2} n - 2 \qquad \text{by algebra,}$$

and so, by the theorem on polynomial orders,

$$2(n-1) + \frac{n(n+1)}{2} + 4\left(\frac{n(n-1)}{2}\right) \text{ is } \Theta(n^2).$$

45. Note that

$$\sum_{k=1}^{n}(k+3) \quad = \quad \sum_{k=1}^{n}k + \sum_{k=1}^{n}3 \qquad \text{by Theorem 5.1.1}$$

$$= \quad \frac{n(n+1)}{2} + \underbrace{(3+3+\cdots+3)}_{n \text{ terms}} \qquad \text{by Theorem 5.2.2}$$

$$= \quad \frac{1}{2}n^2 + \frac{1}{2}n + 3n \qquad \text{by definition of multiplication}$$

$$= \quad \frac{1}{2}n^2 + \frac{7}{2}n \qquad \text{by algebra,}$$

and so, by the theorem on polynomial orders,

$$\sum_{k=1}^{n}(k+3) \ \text{ is } \ \Theta(n^2).$$

48. *a.* <u>Proof:</u> Suppose $a_0, a_1, a_2, \ldots, a_n$ are real numbers and $a_n \neq 0$. Then

$$\lim_{x \to \infty}\left| \frac{a_n x^n + a_{n-1}x^{n-1} + \cdots + a_2 x^2 + a_1 x + a_0}{a_n x^n} \right|$$

$$= \lim_{x \to \infty}\left(1 + \left|\frac{a_{n-1}}{a_n}\right|\frac{1}{x} + \cdots + \left|\frac{a_2}{a_n}\right|\frac{1}{x^{n-2}} + \left|\frac{a_1}{a_n}\right|\frac{1}{x^{n-1}} + \left|\frac{a_0}{a_n}\right|\frac{1}{x^n} \right)$$

$$= \lim_{x \to \infty}1 + \left|\frac{a_{n-1}}{a_n}\right|\lim_{x \to \infty}\left(\frac{1}{x}\right) + \cdots + \left|\frac{a_2}{a_n}\right|\lim_{x \to \infty}\left(\frac{1}{x^{n-2}}\right) + \left|\frac{a_1}{a_n}\right|\lim_{x \to \infty}\left(\frac{1}{x^{n-1}}\right) + \left|\frac{a_0}{a_n}\right|\lim_{x \to \infty}\left(\frac{1}{x^n}\right)$$

$$= 1$$

because $\displaystyle\lim_{x \to \infty}\left(\frac{1}{x^k}\right) = 0$ for all integers $k \geq 1$.

b. <u>Proof:</u>

Suppose $a_0, a_1, a_2, \ldots, a_n$ are real numbers and $a_n \neq 0$. By part (a) and the definition of limit, we can make the following statement: For all positive real numbers ε, there exists a real number M (which we may take to be positive) such that

$$1 - \varepsilon < \left| \frac{a_n x^n + a_{n-1}x^{n-1} + \cdots + a_2 x^2 + a_1 x + a_0}{a_n x^n} \right| < 1 + \varepsilon \quad \text{for all real numbers } x > M.$$

Let $\varepsilon = 1/2$. Then there exists a real number M_0 such that

$$1 - \frac{1}{2} < \left| \frac{a_n x^n + a_{n-1}x^{n-1} + \cdots + a_2 x^2 + a_1 x + a_0}{a_n x^n} \right| < 1 + \frac{1}{2} \quad \text{for all real numbers } x > M_0.$$

Equivalently, for all real numbers $x > M_0$,

$$\frac{1}{2}|a_n|\,|x^n| < \left| a_n x^n + a_{n-1}x^{n-1} + \cdots + a_2 x^2 + a_1 x + a_0 \right| < \frac{3}{2}|a_n|\,|x^n|.$$

Let $A = \frac{1}{2}|a_n|$, $B = \frac{3}{2}|a_n|$, and $k = M_0$. Then

$$A\,|x^n| \leq \left| a_n x^n + a_{n-1}x^{n-1} + \cdots + a_2 x^2 + a_1 x + a_0 \right| \leq B\,|x^n| \quad \text{for all real numbers } x > k,$$

and so, by definition of Θ-notation,

$$a_n x^n + a_{n-1}x^{n-1} + \cdots + a_2 x^2 + a_1 x + a_0 \ \text{ is } \ \Theta(n^n).$$

57. *b.* <u>Proof (by mathematical induction)</u>: Let the property $P(n)$ be the inequality

$$\frac{1}{2}n^{3/2} \leq \sqrt{1} + \sqrt{2} + \sqrt{3} + \cdots + \sqrt{n}. \qquad \leftarrow P(n)$$

Show that $P(1)$ is true: We must show that $\frac{1}{2} \cdot 1^{3/2} \leq \sqrt{1}$. But the left-hand side of the inequality is $1/2$ and the right-hand side is 1, and $1/2 < 1$. So $P(1)$ is true.

Show that for all integers $k \geq 1$, if $P(k)$ is true then $P(k+1)$ is true: Let k be any integer with $k \geq 1$, and suppose that

$$\frac{1}{2}k^{3/2} \leq \sqrt{1} + \sqrt{2} + \sqrt{3} + \cdots + \sqrt{k}. \qquad \begin{array}{l} \leftarrow P(k) \\ \text{inductive hypothesis} \end{array}$$

We must show that

$$\frac{1}{2}(k+1)^{3/2} \leq \sqrt{1} + \sqrt{2} + \sqrt{3} + \cdots + \sqrt{k+1}. \qquad \leftarrow P(k+1)$$

By adding $\sqrt{k+1}$ to both sides of the inductive hypothesis, we have

$$\frac{1}{2}k^{3/2} + \sqrt{k+1} \leq \sqrt{1} + \sqrt{2} + \sqrt{3} + \cdots + \sqrt{k} + \sqrt{k+1}.$$

Thus, by the transitivity of order, it suffices to show that

$$\frac{1}{2}(k+1)^{3/2} \leq \frac{1}{2}k^{3/2} + \sqrt{k+1}.$$

Now when $k \geq 1$,

$$k^2 \geq k^2 - 1 = (k-1)(k+1).$$

Divide both sides by $k(k-1)$ to obtain

$$\frac{k}{k-1} \geq \frac{k+1}{k}.$$

But $\dfrac{k+1}{k} \geq 1$, and any number greater than or equal to 1 is greater than or equal to its own square root. Thus

$$\frac{k}{k-1} \geq \frac{k+1}{k} \geq \sqrt{\frac{k+1}{k}} = \frac{\sqrt{k+1}}{\sqrt{k}}.$$

Hence

$$k\sqrt{k} \geq (k-1)\sqrt{k+1} = (k+1-2)\sqrt{k+1} = (k+1)^{3/2} - 2\sqrt{k+1}.$$

Multiplying the extreme-left and extreme-right sides of the inequality by $1/2$ gives

$$\frac{1}{2}k^{3/2} \geq \frac{1}{2}(k+1)^{3/2} - \sqrt{k+1}, \quad \text{or, equivalently,} \quad \frac{1}{2}(k+1)^{3/2} \leq \frac{1}{2}k^{3/2} + \sqrt{k+1}.$$

[This is what was to be shown].

60. <u>Proof</u>: Suppose $f(x)$ and $g(x)$ are $o(h(x))$ and a and b are any real numbers. Then by properties of limits,

$$\lim_{x \to \infty} \frac{af(x) + bg(x)}{h(x)} = a \lim_{x \to \infty} \frac{f(x)}{h(x)} + b \lim_{x \to \infty} \frac{g(x)}{h(x)} = a \cdot 0 + b \cdot 0 = 0.$$

So $af(x) + bg(x)$ is $o(h(x))$.

Section 11.3

3. *a.* When the input size is increased from m to $2m$, the number of operations increases from cm^3 to $c(2m)^3 = 8cm^3$.

b. By part (a), the number of operations increases by a factor of $\dfrac{8cm^3}{cm^3} = 8$.

c. When the input size is increased by a factor of 10 (from m to $10m$), the number of operations increases by a factor of $\dfrac{c(10m^3)}{cm^3} = \dfrac{1000cm^3}{cm^3} = 1000$.

12. *a.* For each iteration of the inner loop there is one comparison. The number of iterations of the inner loop can be deduced from the following table, which shows the values of k and i for which the inner loop is executed.

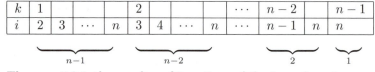

k	1				2				$\cdots$	$n-2$		$n-1$
i	2	3	$\cdots$	n	3	4	$\cdots$	n	$\cdots$	$n-1$	n	n

Therefore, by Theorem 5.2.2, the number of iterations of the inner loop is

$$(n-1) + (n-2) + \cdots + 2 + 1 = \frac{n(n-1)}{2}.$$

It follows that the total number of elementary operations that must be performed when the algorithm is executed is

$$1 \cdot \left(\frac{n(n-1)}{2}\right) = \frac{1}{2}n^2 - \frac{1}{2}n.$$

By the theorem on polynomial orders, $\frac{1}{2}n^2 - \frac{1}{2}n$ is $\Theta(n^2)$, and so the algorithm segment has order n^2.

15. *a.* There are three multiplications for each iteration of the inner loop, and there is one additional addition for each iteration of the outer loop. The number of iterations of the inner loop can be deduced from the following table, which shows the values of i and j for which the inner loop is executed.

i	1				2				$\cdots$	$n-2$		$n-1$
j	2	3	$\cdots$	n	3	4	$\cdots$	n	$\cdots$	$n-1$	n	n

Hence, by Theorem 5.2.2, the total number of iterations of the inner loop is

$$(n-1) + (n-2) + \cdots + 2 + 1 = \frac{n(n-1)}{2}.$$

Because three multiplications are performed for each iteration of the inner loop, the number of operations that are performed when the inner loop is executed is

$$3 \cdot \frac{n(n-1)}{2} = \frac{3}{2}(n^2 - n) = \frac{3}{2}n^2 - \frac{3}{2}n.$$

Now an additional operation is performed each time the outer loop is executed, and because the outer loop is executed n times, this gives an additional n operations. Therefore, the total number of operations is

$$\left(\frac{3}{2}n^2 - \frac{3}{2}n\right) + n = \frac{3}{2}n^2 - \frac{1}{2}n.$$

b. By the theorem on polynomial orders, $\frac{3}{2}n^2 - \frac{1}{2}n$ is $\Theta(n^2)$, and so the algorithm segment has order n^2.

18. *a.* There is one multiplication for each iteration of the inner loop. If n is odd, the number of iterations of the inner loop can be deduced from the following table, which shows the values of i and j for which the inner loop is executed.

i	1				2				$\cdots$	$n-2$					$n-1$					n				
$\left\lfloor\frac{i+1}{2}\right\rfloor$	1				1				$\cdots$	$\frac{n-1}{2}$					$\frac{n-1}{2}$					$\frac{n+1}{2}$				
j	1	2	$\cdots$	n	1	2	$\cdots$	n	$\cdots$	$\frac{n-1}{2}$	$\frac{n-1}{2}+1$	$\cdots$	n		$\frac{n-1}{2}$	$\frac{n-1}{2}+1$	$\cdots$	n		$\frac{n+1}{2}$	$\frac{n+1}{2}+1$	$\cdots$	n	
	$\underbrace{\qquad}_{n}$				$\underbrace{\qquad}_{n}$					$\underbrace{\qquad}_{n-\frac{n-1}{2}+1=\frac{n+3}{2}}$					$\underbrace{\qquad}_{n-\frac{n-1}{2}+1=\frac{n+3}{2}}$					$\underbrace{\qquad}_{n-\frac{n+1}{2}+1=\frac{n+1}{2}}$				

Thus the number of iterations of the inner loop is

$$n + n + (n-1) + (n-1) + \cdots + \frac{n+3}{2} + \frac{n+3}{2} + \frac{n+1}{2}$$

$$= 2\left(n + (n-1) + \cdots + \frac{n+3}{2}\right) + \frac{n+1}{2}$$

$$= 2 \cdot \left(\sum_{k=1}^{n} k - \sum_{k=1}^{(n+1)/2} k\right) + \frac{n+1}{2} \qquad \text{because } \tfrac{n+3}{2} - 1 = \tfrac{n+1}{2}$$

$$= n(n+1) - \frac{n+1}{2}\left(\frac{n+1}{2}+1\right) + \frac{n+1}{2} \qquad \text{by Theorem 5.2.2}$$

$$= \frac{4n(n+1)}{4} - \frac{(n+1)^2}{4}$$

$$= \frac{4n^2 + 4n - n^2 - 2n - 1}{4}$$

$$= \frac{3n^2 + 2n - 1}{4}$$

$$= \frac{3}{4}n^2 + \frac{1}{2}n - \frac{1}{4}.$$

By similar reasoning, if n is even, then the number of iterations of the inner loop is

$$n + n + (n-1) + (n-1) + \cdots + \frac{n+2}{2} + \frac{n+2}{2}$$

$$= 2\left(n + (n-1) + \cdots + \frac{n}{2}\right)$$

$$= 2 \cdot \left(\sum_{k=1}^{n} k - \sum_{k=1}^{n/2} k\right) \qquad \text{because } \tfrac{n+2}{2} - 1 = \tfrac{n}{2}$$

$$= n(n+1) - \frac{n}{2}\left(\frac{n}{2}+1\right) \qquad \text{by Theorem 5.2.2}$$

$$= \frac{4n(n+1)}{4} - \frac{n^2}{4} - \frac{2n}{4}$$

$$= \frac{4n^2 + 4n - n^2 - 2n}{4}$$

$$= \frac{3n^2 + 2n}{4}$$

$$= \frac{3}{4}n^2 + \frac{1}{2}n.$$

Because one operation is performed for each iteration of the inner loop, the answer is that

$$1 \cdot \left(\frac{3}{4}n^2 + n - \frac{3}{4}\right) = \frac{3}{4}n^2 + 3n - \frac{11}{4}$$

elementary operations are performed when n is odd and

$$1 \cdot \left(\frac{3}{4}n^2 + n - 1\right) = \frac{3}{4}n^2 + 6n$$

elementary operations are performed when n is even.

b. By the theorem on polynomial orders, $\frac{3}{4}n^2 + 3n - \frac{11}{4}$ is $\Theta(n^2)$ and $\frac{3}{4}n^2 + 6n$ is also $\Theta(n^2)$ and so this algorithm segment has order n^2.

21.

	$a[1]$	$a[2]$	$a[3]$	$a[4]$	$a[5]$
initial order	7	3	6	9	5
result of step $k = 2$	3	7	6	9	5
result of step $k = 3$	3	6	7	9	5
result of step $k = 4$	3	6	7	9	5
result of step $k = 5$	3	5	6	7	9

27. a.

$$
\begin{aligned}
E_1 &= 0 \\
E_2 &= E_1 + 2 + 1 &= 3 \\
E_3 &= E_2 + 3 + 1 &= 3 + 4 \\
E_4 &= E_3 + 4 + 1 &= 3 + 4 + 5 \\
E_5 &= E_4 + 5 + 1 &= 3 + 4 + 5 + 6
\end{aligned}
$$

$$\vdots$$

Guess:
$$
\begin{aligned}
E_n &= 3 + 4 + 5 + \cdots + (n+1) = [1 + 2 + 3 + 4 + 5 + \cdots + (n+1)] - (1+2) \\
&= \frac{(n+1)(n+2)}{2} - 3 = \frac{n^2 + 3n + 2 - 6}{2} = \frac{n^2 + 3n - 4}{2}
\end{aligned}
$$

b. Proof (by mathematical induction): Let $E_1, E_2, E_3, \ldots$ be a sequence that satisfies the recurrence relation $E_k = E_{k-1} + k + 1$ for all integers $k \geq 2$, with initial condition $E_1 = 0$, and let the property $P(n)$ be the equation

$$E_n = \frac{n^2 + 3n - 4}{2}. \qquad \leftarrow P(n)$$

Show that $P(1)$ is true: The left-hand side of $P(1)$ is E_1, which equals 0 by definition of $E_1, E_2, E_3, \ldots$, and the right-hand side of $P(1)$ is

$$\frac{1^2 + 3 \cdot 1 - 4}{2} = \frac{1 + 3 - 4}{2} = 0$$

also. So $P(1)$ is true.

Show that for all integers $k \geq 1$, if $P(k)$ is true then $P(k+1)$ is true: Let k be any integer with $k \geq 1$, and suppose that

$$E_k = \frac{k^2 + 3k - 4}{2}. \qquad \leftarrow \begin{array}{l} P(k) \\ \text{inductive hypothesis} \end{array}$$

We must show that

$$E_{k+1} = \frac{(k+1)^2 + 3(k+1) - 4}{2}. \qquad \leftarrow P(k+1)$$

But the left-hand side of $P(k+1)$ equals

$$
\begin{aligned}
E_{k+1} &= E_k + (k+1) + 1 &\text{by definition of } E_1, E_2, E_3, \ldots \\
&= \frac{k^2 + 3k - 4}{2} + (k+1) + 1 &\text{by inductive hypothesis} \\
&= \frac{k^2 + 3k - 4 + 2k + 4}{2} \\
&= \frac{k^2 + 5k}{2}.
\end{aligned}
$$

And the right-hand side of $P(k+1)$ equals

$$
\frac{(k+1)^2 + 3(k+1) - 4}{2} = \frac{k^2 + 2k + 1 + 3k + 3 - 4}{2} = \frac{k^2 + 5k}{2}
$$

also, and so $P(k+1)$ is true *[as was to be shown]*.

33. As i goes from $k+1$ to 5 through $5 - (k+1) + 1 = 5 - k$ values (where k goes from 1 to 4), the number of comparisons is

$$
(5-1) + (5-2) + (5-3) + (5-4) = 4 + 3 + 2 + 1 = 10.
$$

39. By the result of exercise 38, $s_n = \dfrac{1}{2}n^2 + \dfrac{3}{2}n$, which is $\Theta(n^2)$ by the theorem on polynomial orders.

42. There are two operations (one addition and one multiplication) per iteration of the loop, and there are n iterations of the loop. Therefore, $t_n = 2n$.

Section 11.4

6.

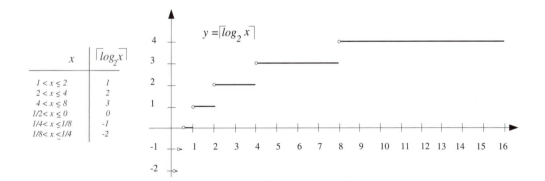

x	$\lceil log_2 x \rceil$
$1 < x \le 2$	1
$2 < x \le 4$	2
$4 < x \le 8$	3
$1/2 < x \le 0$	0
$1/4 < x \le 1/8$	-1
$1/8 < x \le 1/4$	-2

12. When $\frac{1}{2} < x < 1$, then $-1 < \log_2 x < 0$.

When $\frac{1}{4} < x < \frac{1}{2}$, then $-2 < \log_2 x < -1$.

When $\frac{1}{8} < x < \frac{1}{4}$, then $-3 < \log_2 x < -2$.

And so forth.

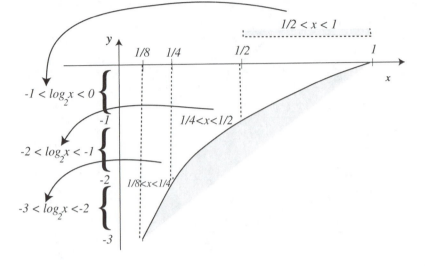

24. <u>Proof (by strong mathematical induction)</u>: Let $c_1, c_2, c_3, \ldots$. be a sequence that satisfies the recurrence relation

$$c_k = 2c_{\lfloor k/2 \rfloor} + k \text{ for all integers } k \geq 2, \text{ with initial condition } c_1 = 0,$$

and let the property $P(n)$ be the inequality

$$c_n \leq n \log_2 n. \qquad \leftarrow P(n)$$

Show that $P(1)$ is true: For $n = 1$ the inequality states that $c_1 \leq 1 \cdot \log_2 1 = 1 \cdot 0 = 0$, which is true because $c_1 = 0$. So $P(1)$ is true.

Show that if $k \geq 1$ and $P(i)$ is true for all integers i from 1 through k, then $P(k + 1)$ is true: Let k be any integer with $k \geq 1$, and suppose that

$$c_i \leq i \log_2 i \text{ for all integers } i \text{ with } 1 \leq i \leq k. \quad \leftarrow \begin{array}{l} \text{inductive} \\ \text{hypothesis} \end{array}$$

We must show that

$$c_{k+1} \leq (k + 1) \log_2(k + 1).$$

First note that because k is greater than 1 and by definition of floor,

$$1 \leq \left\lfloor \frac{k + 1}{2} \right\rfloor \leq \frac{k + 1}{2}.$$

Also, because k is an integer with $k \geq 1$, we have

$$1 \leq k \Rightarrow k + 1 \leq k + k \Rightarrow k + 1 \leq 2k \Rightarrow \frac{k + 1}{2} \leq k.$$

Thus, by the transitive property of order,

$$\left\lfloor \frac{k + 1}{2} \right\rfloor \leq k.$$

Then

$$c_{k+1} = 2c_{\lfloor (k+1)/2 \rfloor} + (k+1)$$

by definition of $c_1, c_2, c_3, \ldots$

$$\leq 2\left(\left\lfloor \frac{k+1}{2} \right\rfloor \log_2 \left\lfloor \frac{k+1}{2} \right\rfloor \right) + (k+1)$$

by inductive hypothesis because $\left\lfloor \frac{k+1}{2} \right\rfloor \leq k$

$$\leq (k+1) \log_2 \left(\frac{k+1}{2} \right) + (k+1)$$

since $1 \leq \left\lfloor \frac{k+1}{2} \right\rfloor \leq \frac{k+1}{2}$, we have by property (11.4.1)
that $\log_2 \left\lfloor \frac{k+1}{2} \right\rfloor \leq \log_2 \left(\frac{k+1}{2} \right)$

$$= (k+1) \left[\log_2(k+1) - \log_2 2 \right] + (k+1)$$

by Theorem 7.2.1(b)

$$= (k+1) \left[\log_2(k+1) - 1 \right] + (k+1)$$

because $\log_2 2 = 1$

$$= (k+1) \log_2(k+1)$$

by algebra,

Therefore, by transitivity of equality and order, $c_{k+1} \leq (k+1) \log_2(k+1)$ *[as was to be shown]*.

33. For all integers $n > 0$,

$$2^n \leq 2^{n+1} \leq 2 \cdot 2^n.$$

Thus, let $A = 1$, $B = 2$, and $k = 0$. Then

$$A \cdot 2^n \leq 2^{n+1} \leq B \cdot 2^n \quad \text{for all integers } n > k,$$

and so, by definition of Θ-notation, 2^{n+1} is $\Theta(2^n)$.

36. By factoring out a 4 and using the formula for the sum of a geometric sequence (Theorem 5.2.3), we have that for all integers $n > 1$,

$$\begin{aligned} 4 + 4^2 + 4^3 + \cdots + 4^n &= 4(1 + 4 + 4^2 + \cdots + 4^{n-1}) \\ &= 4 \left(\frac{4^{(n-1)+1} - 1}{4 - 1} \right) \\ &= \frac{4}{3}(4^n - 1) \\ &= \frac{4}{3} \cdot 4^n - \frac{4}{3} \\ &\leq \frac{4}{3} \cdot 4^n. \end{aligned}$$

Moreover, because

$$4 + 4^2 + 4^3 + \cdots + 4^{n-1} \geq 0, \quad \text{then} \quad 4^n \leq 4 + 4^2 + 4^3 + \cdots + 4^{n-1} + 4^n.$$

So let $A = 1$, $B = 4/3$, and $k = 1$. Then, because all quantities are positive,

$$A \cdot |4^n| \leq \left| 4 + 4^2 + 4^3 + \cdots + 4^n \right| \leq B \cdot |4^n| \quad \text{for all integers } n > k,$$

and thus, by definition of Θ-notation, $4 + 4^2 + 4^3 + \cdots + 4^n$ is $\Theta(4^n)$.

42. $1 + \dfrac{1}{2} = \dfrac{3}{2}, \quad 1 + \dfrac{1}{2} + \dfrac{1}{3} = \dfrac{11}{6}, \quad 1 + \dfrac{1}{2} + \dfrac{1}{3} + \dfrac{1}{4} = \dfrac{50}{24} = \dfrac{25}{12}, \quad 1 + \dfrac{1}{2} + \dfrac{1}{3} + \dfrac{1}{4} + \dfrac{1}{5} = \dfrac{137}{60}$

45. *a.* <u>Proof</u>: If n is any positive integer, then $\log_2 n$ is defined and by definition of floor,

$$\lfloor \log_2 n \rfloor \ \leq \ \log_2 n \ < \ \lfloor \log_2 n \rfloor + 1.$$

If, in addition, n is greater than 2, then since the logarithmic function with base 2 is increasing

$$\log_2 n \ > \ \log_2 2 \ = \ 1.$$

Thus, by definition of floor,

$$1 \ \leq \ \lfloor \log_2 n \rfloor .$$

Adding $\lfloor \log_2 n \rfloor$ to both sides of this inequality gives

$$\lfloor \log_2 n \rfloor + 1 \ \leq \ 2 \lfloor \log_2 n \rfloor .$$

Hence, by the transitive property of order (T18 in Appendix A),

$$\log_2 n \ \leq \ 2 \lfloor \log_2 n \rfloor ,$$

and dividing both sides by 2 gives

$$\frac{1}{2} \log_2 n \ \leq \ \lfloor \log_2 n \rfloor .$$

Let $A = 1/2$, $B = 1$, and $k = 2$. Then

$$A \log_2 n \ \leq \ \lfloor \log_2 n \rfloor \ \leq \ B \log_2 n \ \text{ for all integers } n \geq k,$$

and, because $\log_2 n$ is positive for $n > 2$, we may write

$$A \left| \log_2 n \right| \ \leq \ \left| \lfloor \log_2 n \rfloor \right| \ \leq \ B \left| \log_2 n \right| \ \text{ for all integers } n \geq k.$$

Therefore, by definition of Θ-notation, $\lfloor \log_2 n \rfloor$ is $\Theta(\log_2 n)$.

b. <u>Proof</u>: If n is any positive real number, then $\log_2 n$ is defined and by definition of floor,

$$\lfloor \log_2 n \rfloor \leq \log_2 n .$$

If, in addition, n is greater than 2, then, as in part (a),

$$\log_2 n \ < \ \lfloor \log_2 n \rfloor + 1 \quad \text{and} \quad \lfloor \log_2 n \rfloor + 1 \ \leq \ 2 \log_2 n .$$

Hence, because $\log_2 n$ is positive for $n > 2$, we may write

$$\left| \log_2 n \right| \leq \left| \lfloor \log_2 n \rfloor + 1 \right| \leq 2 \left| \log_2 n \right| .$$

Let $A = 1$, $B = 2$ and $k = 2$. Then

$$A \left| \log_2 n \right| \ \leq \ \left| \lfloor \log_2 n \rfloor + 1 \right| \ \leq \ B \left| \log_2 n \right| \ \text{ for all integers } n \geq k.$$

Therefore, by definition of Θ-notation, $\lfloor \log_2 n \rfloor + 1$ is $\Theta(\log_2 n)$.

48. <u>Proof</u>:

Suppose n is a variable that takes positive integer values. Then whenever $n \geq 2$,

$$2^n \ = \ \underbrace{2 \cdot 2 \cdot 2 \cdot 2 \cdot 2 \cdots 2}_{n \text{ factors}} \ \leq \ \underbrace{2 \cdot 2 \cdot 3 \cdot 4 \cdot 5 \cdots n}_{n \text{ factors}} \ \leq \ 2n!.$$

Let $B = 2$ and $b = 2$. Since 2^n and $n!$ are positive for all n,

$$\left| 2^n \right| \ \leq \ B |n!| \ \text{ for all integers } n \geq b.$$

Hence by definition of O-notation 2^n is $O(n!)$.

51. *a.* Let n be any positive integer. Then for any real number x *[because $u < 2^u$ for all real numbers u]*,

$$\frac{x}{n} < 2^{\frac{x}{n}} \quad \Rightarrow \quad x < n2^{\frac{x}{n}} \quad \Rightarrow \quad x^n < (n2^{\frac{x}{n}})^n = n^n \cdot 2^x.$$

So $x^n < n^n 2^x$.

b. Let x be any positive real number and let n be any positive integer. Then

$$x^n = |x^n| \quad \text{and} \quad n^n 2^x = 2^x \, |n^n| \,,$$

and thus the result of part (a) may be written as

$$|x^n| \le 2^x \, |n^n| \,.$$

Let $B = 2^x$ and $b = 0$. Then $|x^n| \le B \, |n^n|$ for all integers $n > b$, and so by definition of O-notation x^n is $O(n^n)$.

Section 11.5

6. *a.*

index	0			
bot	1	1	1	1
top	10	4	1	0
mid		5	2	1

b.

index	0		8
bot	1	6	
top	10		
mid		5	8

12.

n	424	141	47	15	5	1	0

15. If $n \ge 3$, then

$$
\begin{array}{rcll}
b_n & = & 1 + \lfloor \log_3 n \rfloor & \text{by the result of exercise 14} \\
\Rightarrow \quad b_n & \le & 1 + \log_3 n & \text{because } \lfloor \log_3 n \rfloor \le \log_3 n \text{ by definition of floor} \\
\Rightarrow \quad b_n & \le & \log_3 n + \log_3 n & \text{because if } n \ge 3 \text{ then } \log_3 n \ge 1 \\
\Rightarrow \quad b_n & \le & 2 \log_3 n & \text{by algebra.}
\end{array}
$$

Furthermore, because $\log_3 n \ge 0$ for $n > 2$, we may write

$$|\log_3 n| < |\lfloor \log_3 n \rfloor + 1| \le 2 \, |\log_3 n| \,.$$

Let $A = 1$, $B = 2$, and $k = 2$. Then all quantities are positive, and so

$$A \, |\log_3 n| < |\lfloor \log_3 n \rfloor + 1| \le B \, |\log_3 n| \quad \text{for all integers } n > k.$$

Hence by definition of Θ-notation, $b_n = 1 + \lfloor \log_3 n \rfloor$ is $\Theta(\log_3 n)$, and thus the algorithm segment has order $\log_3 n$.

18. Suppose an array of length k is input to the **while** loop and the loop is iterated one time. The elements of the array can be matched with the integers from 1 to k with $m = \left\lceil \dfrac{k+1}{2} \right\rceil$, as shown below:

Case 1 (k is even): In this case $m = \left\lceil \dfrac{k+1}{2} \right\rceil = \left\lceil \dfrac{k}{2} + \dfrac{1}{2} \right\rceil = \dfrac{k}{2} + 1$, and so the number of elements in the left subarray equals $m - 1 = (\dfrac{k}{2} + 1) - 1 = \dfrac{k}{2} = \left\lfloor \dfrac{k}{2} \right\rfloor$. The number of elements in the right subarray equals $k - (m+1) - 1 = k - m = k - (\dfrac{k}{2} + 1) = \dfrac{k}{2} - 1 < \left\lfloor \dfrac{k}{2} \right\rfloor$. Hence both subarrays (and thus the new input array) have length at most $\left\lfloor \dfrac{k}{2} \right\rfloor$.

Case 2 (k is odd): In this case $m = \left\lceil \dfrac{k+1}{2} \right\rceil = \dfrac{k+1}{2}$, and so the number of elements in the left subarray equals $m - 1 = \dfrac{k+1}{2} - 1 = \dfrac{k-1}{2} = \left\lfloor \dfrac{k}{2} \right\rfloor$. The number of elements in the right subarray equals $k - m = k - \dfrac{k+1}{2} = \dfrac{k-1}{2} = \left\lfloor \dfrac{k}{2} \right\rfloor$ also. Hence both subarrays (and thus the new input array) have length $\left\lfloor \dfrac{k}{2} \right\rfloor$.

The arguments in cases 1 and 2 show that the length of the new input array to the next iteration of the **while** loop has length at most $\lfloor k/2 \rfloor$.

21.

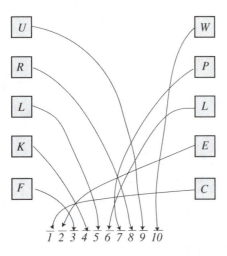

24. *a.* Refer to Figure 11.5.3. Observe that when k is odd, the subarray $a[mid+1], a[mid+2], \ldots a[top]$ has length
$$k - \left(\dfrac{k+1}{2} + 1 \right) + 1 = \dfrac{k-1}{2} = \left\lfloor \dfrac{k}{2} \right\rfloor .$$

And when k is even, the subarray $a[mid + 1], a[mid + 2], \ldots a[top]$ has length
$$k - \left(\dfrac{k}{2} + 1 \right) + 1 = \dfrac{k}{2} = \left\lfloor \dfrac{k}{2} \right\rfloor .$$

So in either case the subarray has length $\lfloor k/2 \rfloor$.

Review Guide: Chapter 11

Definitions: How are the following terms defined?

- real-valued function of a real variable *(p. 717)*
- graph of a real-valued function of a real variable *(p. 717)*
- power function with exponent a *(p. 718)*
- floor function *(p. 719)*
- multiple of a real-valued function of a real variable *(p. 721)*
- increasing function *(p. 722)*
- decreasing function *(p. 722)*
- $f(x)$ is $\Omega(g(x)$, where f and g are real-valued functions of a real variable defined on the same set of nonnegative real numbers *(p. 727)*
- $f(x)$ is $O(g(x)$, where f and g are real-valued functions of a real variable defined on the same set of nonnegative real numbers *(p. 727)*
- $f(x)$ is $\Theta(g(x)$, where f and g are real-valued functions of a real variable defined on the same set of nonnegative real numbers *(p. 727)*
- algorithm A is $\Theta(g(n))$ (or A has order $g(n)$) *(p. 741)*
- algorithm A is $\Omega(g(n))$ (or A has a best case order $g(n)$) *(p. 741)*
- algorithm A is $O(g(n))$ (or A has a worst case order $g(n)$) *(p. 741)*
- polynomial time algorithms, NP class, NP-complete problems, the P vs. NP problem, tractable and intractable problems *(pp. 775-776)*

Polynomial and Rational Functions and Their Orders

- What is the graph of the floor function? *(pp. 719-720)*
- What is the difference between the graph of a function defined on an interval of real numbers and the graph of a function defined on a set of integers? *(p. 720)*
- How do you graph a multiple of a real-valued function of a real variable? *(p. 721)*
- How do you prove that a function is increasing (decreasing)? *(p. 723)*
- What are some properties of O-, Ω-, and Θ-notation? Can you prove them? *(p. 728)*
- If $x > 1$, what is the relationship between x^r and x^s, where r and s are rational numbers and $r < s$? *(p. 729)*
- Given a polynomial, how do you use the definition of Θ-notation to show that the polynomial has order x^n, where n is the degree of the polynomial? *(pp. 730-732)*
- What is the theorem on polynomial orders? *(p. 733)*
- What is an order for the sum of the first n integers? *(p. 735)*
- What is an order for a function that is a ratio of rational power functions? *(p. 736)*

Efficiency of Algorithms

- How do you compute the order of an algorithm segment that contains a loop? a nested loop? *(pp. 742-744)*
- How do you find the number of times a loop will iterate when an algorithm segment is executed? *(p. 743)*
- How do you use the theorem on polynomial orders to help find the order of an algorithm segment? *(p. 744)*
- What is the sequential search algorithm? How do you compute its worst case order? its average case order? *(pp. 739-740)*
- What is the insertion sort algorithm? How do you compute its best and worst case orders? *(pp. 740, 744-746)*

Logarithmic and Exponential Orders

- What do the graphs of logarithmic and exponential functions look like? *(pp. 751-752)*
- What can you say about the base 2 logarithm of a number that is between two consecutive powers of 2? *(p. 753)*
- How do you compute the number of bits needed to represent a positive integer in binary notation? *(p. 755)*
- How are logarithms used to solve recurrence relations? *(pp. 755-757)*
- If $b > 1$, what can you say about the relation among $\log_b x$, x^r, and $x \log_b x$? *(p. 758)*
- If $b > 1$ and $c > 1$, how are orders of $\log_b x$ and $\log_c x$ related? *(p. 760)*
- What is an order for a harmonic sum? *(pp. 760-762)*
- What is a divide-and-conquer algorithm? *(p. 765)*
- What is the binary search algorithm? *(pp. 765-767)*
- What is the worst case order for the binary search algorithm, and how do you find it? *(pp. 768-772)*
- What is the merge sort algorithm? *(pp. 772-775)*
- What is the worst case order for the merge sort algorithm, and how do you find it? *(p. 775)*

Chapter 12: Regular Expressions and Finite-State Automata

This chapter opens with some historical background about the connections between computers and formal languages. Section 12.1 focuses on regular expressions and emphasizes their utility for pattern matching, whether for compilers or for general text processing.

Section 12.2 introduces the concept of finite-state automaton. In one sense, it is a natural sequel to the discussions of digital logic circuits in Section 2.4 and Boolean functions in Section 7.1, with the next-state function of an automaton governing the operation of sequential circuit in much the same way that a Boolean function governs the operation of a combinatorial circuit. The section also provides practice in finding a finite-state automaton that corresponds to a regular expression and shows how to write a program to implement a finite-state automaton. Both abilities are useful for computer programming. The section ends with a statement and partial proof of Kleene's theorem, which describes the exact nature of the relationship between finite-state automata and regular languages.

The equivalence and simplification of finite-state automata, discussed in Section 12.3, provides an additional application for the concept of equivalence relation, introduced in Section 8.3. Note the parallel between the simplification of digital logic circuits discussed in Section 2.4 and the simplification of finite-state automata developed in this section. Both kinds of simplification have obvious practical use.

Section 12.1

3. b. $L = \{11*, 11/, 12*, 12/, 21*, 21/, 22*, 22/\}$

   $11* = 1 * 1 = 1$, $11/ = 1/1 = 1$, $12* = 1 * 2 = 2$, $12/ = 1/2 = 0.5$, $21* = 2 * 1 = 2$, $21/ = 2/1 = 2$, $22* = 2 * 2 = 4$, $22/ = 2/2 = 1$

6. $L_1 L_2$ is the set of strings of 0's and 1's that both start and end with a 0.

   $L_1 \cup L_2$ is the set of strings of 0's and 1's that start with a 0 or end with a 0 (or both).

   $(L_1 \cup L_2)^*$ is the set of strings of 0's and 1's that start with a 0 or end with a 0 (or both) or that contain 00.

9. $(((x \mid (y(z^*)))^*)((yx) \mid (((yz)^*)z)))$

12. $xy(x^*y)^* \mid (yx \mid y)y^*$

15. $L((a \mid b)c) = L(a \mid b)L(c) = (L(a) \cup L(b))L(c) = (\{a\} \cup \{b\})\{c\} = \{a, b\})\{c\} = \{ac, bc\}$

18. $x, yxxy, xx, xyxxy, xyxxyyxxy, \ldots$

21. The language consists of the set of all strings of x's and y's that start with xy or yy followed by any string of x's and y's.

24. The string 120 does not belong to the language defined by $(01^*2)^*$ because it does not start with 0. However, 01202 does belong to the language because 012 and 02 are both defined by 01^*2 and the language is closed under concatenation.

27. $x \mid y^* \mid y^*(xyy^*)(\epsilon \mid x)$

30. Note that for any regular expression x, $(x^*)^*$ defines the set of all strings obtained by concatenating a finite number of a finite number of concatenations of copies of x. But any such string can equally well be obtained simply by concatenating a finite number of copies of x, and thus $(x^*)^* = x^*$. Hence the given languages are the same: $L((rs)^*) = L(((rs)^*)^*)$.

33. $[a-z]\{3\}[a-z]^*ly$

187

Section 12.2

3. *a.* U_0, U_1, U_2, U_3 *b.* a, b *c.* U_0 *d.* U_3

e.

	state	input a	input b
→	U_0	U_2	U_1
	U_1	U_2	U_3
	U_2	U_2	U_2
◎	U_3	U_3	U_3

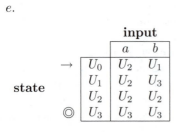

6. *a.* s_0, s_1, s_2, s_3 *b.* $0, 1$ *c.* s_0 *d.* s_0

e.

		state	input 0	input 1
→	◎	s_0	s_0	s_1
		s_1	s_1	s_2
		s_2	s_2	s_3
		s_3	s_3	s_0

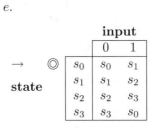

9. *a.* s_0, s_1, s_2, s_3 *b.* $0, 1$ *c.* s_0 *d.* s_1

e.

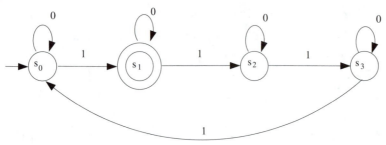

21. *a.*

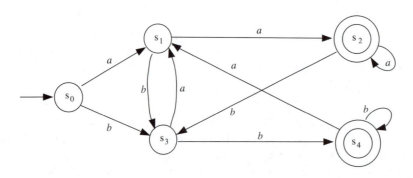

b. $(a|b)^*(aa|bb)$

24. *a.*

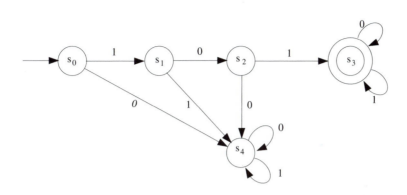

b. $101(0|1)^*$

27. *a.*

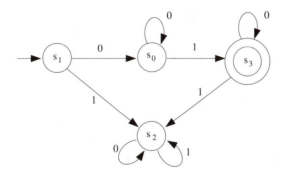

b. 00^*10^* (or using the $^+$ notation: 0^+10^*)

30.

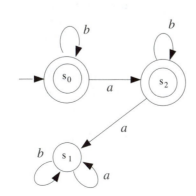

48. Let d represent the character class $[0-9]$.

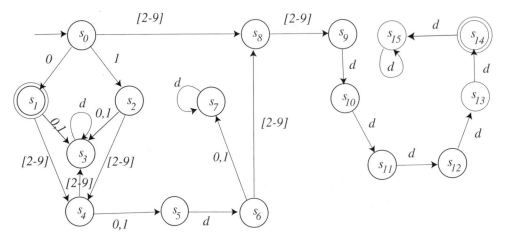

51. Proof (by contradiction):

Suppose there were a finite-state automaton A that accepts L. Consider all strings of the form a^i for some integer $i \geq 0$.

Since the set of all such strings is infinite and the number of states of A is finite, by the pigeonhole principle at least two of these strings, say a^p and a^q with $p < q$, must send A to the same state, say s, when input to A starting in its initial state. (The strings of the given form are the pigeons, the states are the pigeonholes, and each string is associated with the state to which A goes when the string is input to A starting in its initial state.)

Because A accepts L, A accepts $a^q b^q$ but A does not accept $a^p b^q$.

Now since $a^q b^q$ is accepted by A, A goes to an accepting state if, starting from the initial state, first a^q is input to it (sending it to state s) and then b^q is input to it. But A also goes to state s after a^p is input to it. Hence, inputting b^q to A after inputting a^p also sends A to an accepting state. In other words, A accepts $a^p b^q$.

Thus $a^p b^q$ is accepted by A and yet it is not accepted by A, which is a contradiction. Hence the supposition is false: there is no finite-state automaton that accepts L.

54. *a.* Proof:

Suppose A is a finite-state automaton with input alphabet $\sum$, and suppose $L(A)$ is the language accepted by A.

Define a new automaton A' as follows: Both the states and the input symbols of A' are the same as the states and input symbols of A. The only difference between A and A' is that each accepting state of A is a non-accepting state of A', and each non-accepting state of A is an accepting state of A'.

It follows that each string in $\sum^*$ that is accepted by A is not accepted by A', and each string in $\sum^*$ that is not accepted by A is accepted by A'. Thus $L(A') = (L(A))^c$.

b. Proof:

Let A_1 and A_2 be finite-state automata, and let $L(A_1)$ and $L(A_2)$ be the languages accepted by A_1 and A_2, respectively.

By part (a), there exist automata A_1' and A_2' such that $L(A_1') = (L(A_1))^c$ and $L(A_2') = (L(A_2))^c$.

Hence, by Kleene's theorem (part 1), there are regular expressions r_1 and r_2 that define $(L(A_1))^c$ and $(L(A_2))^c$, respectively. So we may write $(L(A_1))^c = L(r_1)$ and $(L(A_2))^c = L(r_2)$.

Now by definition of regular expression, $r_1 \mid r_2$ is a regular expression, and, by definition of the language defined by a regular expression, $L(r_1 \mid r_2) = L(r_1) \cup L(r_2)$.

Thus, by substitution and De Morgan's law, $L(r_1 \mid r_2) = (L(A_1))^c \cup (L(A_2))^c = (L(A_1) \cap L(A_2))^c$, and so, by Kleene's theorem (part(2)), there is a finite-state automaton, say A, that accepts $(L(A_1) \cap L(A_2))^c$.

It follows from part (a) that there is a finite-state automaton, A', that accepts $((L(A_1) \cap L(A_2))^c)^c$. But, by the double complement law for sets, $((L(A_1) \cap L(A_2))^c)^c = L(A_1) \cap L(A_2)$.

So there is a finite-state automaton, A', that accepts $L(A_1) \cap L(A_2)$, and hence, by Kleene's theorem and the definition of regular language, $L(A_1) \cap L(A_2)$ is a regular language.

Section 12.3

3. *a.* 0-equivalence classes: $\{s_1, s_3\}$, $\{s_0, s_2\}$

   1-equivalence classes: $\{s_1, s_3\}$, $\{s_0, s_2\}$

   b. transition diagram for $\bar{A}$:

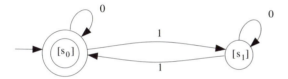

6. *a.* 0-equivalence classes: $\{s_0, s_1, s_3, s_4, s_5\}$, $\{s_2, s_6\}$

   1-equivalence classes: $\{s_0, s_4, s_5\}$, $\{s_1, s_3\}$, $\{s_2\}$, $\{s_6\}$

   2-equivalence classes: $\{s_0, s_4\}$, $\{s_5\}$, $\{s_1\}$, $\{s_3\}$, $\{s_2\}$, $\{s_6\}$

   3-equivalence classes: $\{s_0\}$, $\{s_4\}$, $\{s_5\}$, $\{s_1\}$, $\{s_3\}$, $\{s_2\}$, $\{s_6\}$

   b. The transition diagram for $\overline{A}$ is the same as the one given for A except that the states are denoted $[s_0], [s_1], [s_2], [s_3], [s_4], [s_5], [s_6]$.

15. <u>Proof:</u>

Suppose k is an integer such that $k \geq 1$ and C_k is a k-equivalence class. We must show that there is a $k-1$ equivalence class, C_{k-1}, such that $C_k \subseteq C_{k-1}$.

By property (12.3.3), the $(k-1)$-equivalence classes partition the set of all states of A in to a union of mutually disjoint subsets.

Let s be any state in C_k. Then s is in *some* $(k-1)$-equivalence class; call it C_{k-1}.

Let t be any other state in C_k. *[We will show that $t \in C_{k-1}$ also.]* Then $t\ R_k\ s$, and so for all input strings of length k, $N^*(t, w)$ is an accepting state $\Leftrightarrow N^*(s, w)$ is an accepting state.

Since $k - 1 < k$, it follows that for all input strings of length $k - 1$, $N^*(t, w)$ is an accepting state $\Leftrightarrow N^*(s, w)$ is an accepting state.

Consequently, $t\ R_{k-1}\ s$, and so t and s are in the same $(k-1)$-equivalence class.

But $s \in C_{k-1}$. Hence $t \in C_{k-1}$ also. We, therefore, conclude that $C_k \subseteq C_{k-1}$.

18. <u>Proof:</u>

Suppose A is an automaton and C is a $*$-equivalence class of states of A.

By Theorem 12.3.2, there is an integer $K \geq 0$ such that C is a K-equivalence class of A. Suppose C contains both an accepting state s and a nonaccepting state t of A.

Since both s and t are in the same K-equivalence class, s is K-equivalent to t (by exercises 36 and 37 of Section 8.3), and so by exercise 17, s is 0-equivalent to t.

But this is impossible because there are only two 0-equivalence classes, the set of all accepting states and the set of all nonaccepting states, and these two sets are disjoint.

Hence the supposition that C contains both an accepting and a nonaccepting state is false: C consists entirely of accepting states or entirely of nonaccepting states.

Review Guide: Chapter 12

Definitions: How are the following terms defined?

- alphabet, string over an alphabet, formal language over an alphabet *(p. 781)*
- Σ^n, Σ^* (the Kleene closure of Σ), and Σ^+ (the positive closure of Σ), where Σ is an alphabet *(p. 781)*
- concatenation of x and y, where x and y are strings *(p. 783)*
- concatenation of L and L', where L and L' are languages *(p. 783)*
- union of L and L', where L and L' are languages *(p. 783)*
- Kleene closure of L , where L is a language *(p. 783)*
- regular expression over an alphabet *(p. 783)*
- language defined by a regular expression *(p. 784)*
- character class *(p. 787)*
- finite-state automaton, next-state function *(p. 793)*
- language accepted by a finite-state automaton *(p. 795)*
- eventual-state function for a finite-state automaton *(p. 797)*
- regular language *(p. 804)*
- $*$-equivalence of states in a finite-state automaton *(p. 809)*
- k-equivalence of states in a finite-state automaton *(p. 810)*
- quotient automaton *(p. 814)*
- equivalent automata *(p. 816)*

Regular Expressions

- What is the order of precedence for the operations in a regular expression? *(p. 784)*
- How do you find the language defined by a regular expression? *(p. 785)*
- Given a language, how do you find a regular expression that defines the language? *(p. 786)*
- What are some practical uses of regular expressions? *(pp. 787-789)*

Finite-State Automata

- How do you construct an annotated next-state table for a finite-state automaton given the transition diagram for the automaton? *(p. 794)*
- How do you construct a transition diagram for a finite-state automaton given its next-state table? *(pp. 794-795)*
- How do you find the state to which a finite-state automaton goes if the characters of a string are input to it? *(p. 796)*
- How do you find the language accepted by a finite-state automaton? *(p. 796)*
- Given a simple formal language, how do you construct a finite-state automaton to accept the language? *(p. 798)*
- How can you use software to simulate the action of a finite-state automaton? *(pp. 799-801)*
- What do the two parts of Kleene's theorem say about the relation between the language accepted by a finite-state automaton and the language defined by a regular expression? *(pp. 799. 803)*
- How can the pigeonhole principle be used to show that a language is not regular? *(p. 804)*
- How do you find the k-equivalence classes for a finite-state automaton? *(p. 811)*
- How do you find the $*$-equivalence classes for a finite-state automaton? *(p. 812)*
- How do you construct the quotient automaton for a finite-state automaton? *(pp. 814-815)*
- What is the relation between the language accepted by a finite-state automaton and the language accepted by the corresponding quotient automaton? *(p. 814)*

Conventions for Mathematical Writing

1. When introducing a new variable into a discussion, the convention is to place the new variable to the left of the equal sign and the expression that defines it to the right. This convention is identical to the one used in computer programming. For example, in a computer program, if a and b have previously been defined, and you want to assign the value of $a + b$ to a new variable s, you would write something like

$$s := a + b.$$

   Similarly, in a mathematical proof, if a and b have previously been introduced into a discussion, and you want to let s be their sum,

   instead of writing "Let $a + b = s$," you should write, "Let $s = a + b$."

2. It is considered good mathematical writing to avoid starting a sentence with a variable. That is one reason that mathematical writing frequently uses words and phrases such as Then, Thus, So, Therefore, It follows that, Hence, etc. For example, in a proof that any sum of even integers is even, instead of writing,

   By definition of even, $m = 2a$ and $n = 2b$ for some integers a and b.

   $$m + n = 2a + 2b \ldots$$

   write

   By definition of even, $m = 2a$ and $n = 2b$ for some integers a and b.
   Then

   $$m + n = 2a + 2b \ldots$$

   The fact that $m + n = 2a + 2b$ is a consequence of the facts that $m = 2a$ and $n = 2b$. Including the word "Then" in your proof alerts your reader to this reasoning.

3. Standard mathematical writing avoids repeating the left-hand side in a sequence of equations in which the left-hand side remains constant. For example, if $n = 5q + 4$, instead of writing

$$
\begin{aligned}
n^2 &= (5q + 4)^2 \\
n^2 &= 25q^2 + 40q + 16 \\
n^2 &= 25q^2 + 40q + 15 + 1 \\
n^2 &= 5(5q^2 + 8q + 3) + 1
\end{aligned}
$$

   all the n^2 except the first are omitted and each subsequent equal sign is read as "which equals," as shown below:

$$
\begin{aligned}
n^2 &= (5q + 4)^2 \\
&= 25q^2 + 40q + 16 \\
&= 25q^2 + 40q + 15 + 1 \\
&= 5(5q^2 + 8q + 3) + 1
\end{aligned}
$$

4. Respecting the equal sign is one of the most important mathematical conventions. An equal sign should only be used between quantities that are equal, not as a substitute for words like "is," "means that," "if and only if," $\Leftrightarrow$, or "is equivalent to." For example, if $a = 4$ and $b = 12$, students occasionally write:

$$a \mid b = 4 \mid 12 \text{ since } 12 = 4 \cdot 3.$$

But if this were read out loud, it would be, "a divides b equals 4 divides 12 since 12 equals 4 times 3," which makes no sense. A correct version would be

$$a \mid b \Leftrightarrow 4 \mid 12, \text{ which is true because } 12 = 4 \cdot 3$$

or

$$a \mid b \text{ because } 4 \mid 12 \text{ since } 12 = 4 \cdot 3.$$

5. It is unnecessary, and even risky, to place full statements of definitions and theorems inside the bodies of proofs. The reason is that the variables used to express them can become confused with variables that are part of the proof. So instead of including the statement of the definition of divisibility, for example, just write, "by definition of divisibility." Similarly, instead of including the statement of, say, Theorem 8.4.3, just write, "by Theorem 8.4.3." For instance, to prove that a sum of any even integer plus any odd integer is odd, someone might write the following:

> Suppose m is any even integer and n is any odd integer.
> For an integer to be even means that it equals $2k$ for some integer k, and
> for an integer to be odd, means that it equals $2k + 1$ for some integer k.
> Thus $m + n = 2k + (2k + 1) = 4k + 1...$

The problem is that although the letter k appears in the statements of the definitions in the text, it refers to a different quantity in each one. However, when the statements are combined together in the proof, the letter k can have only one interpretation. The result is that the argument in the "proof" only applies to an even integer and the next successive odd integer, not to *any* even integer and *any* odd integer.

Tips for Success with Proofs and Disproofs

Make sure your proofs are genuinely convincing. Express yourself carefully and completely – but concisely! Write in complete sentences, but don't use an unnecessary number of words.

Disproof by Counterexample

- To disprove a universal statement, give a counterexample.
- Write the word "Counterexample" at the beginning of a counterexample.
- Write counterexamples in complete sentences.
- Give values of the variables that you believe show the property is false.
- Include the computations that prove beyond any doubt that these values really do make the property false.

All Proofs

- Write the word "Proof" at the beginning of a proof.
- Write proofs in complete sentences.
- Start each sentence with a capital letter and finish with a period.

Direct Proof

- Begin each direct proof with the word "Suppose."
- In the "Suppose" sentence:
  - Introduce a variable or variables (indicating the general set they belong to - e.g., integers, real numbers etc.), and
  - Include the hypothesis that the variables satisfy.
- Identify the conclusion that you will need to show in order to complete the proof.
- Reason carefully from the "suppose" to the "conclusion to be shown."
- Include the little words (like "Then," "Thus," "So," "It follows that") that make your reasoning clear.
- Give a reason to support each assertion you make in your proof.

Proof by Contradiction

- Begin each proof by contradiction by writing "Suppose not. That is, suppose...," and continue this sentence by carefully writing the negation of the statement to be proved.
- After you have written the "suppose," you need to show that this supposition leads logically to a contradiction.
- Once you have derived a contradiction, you can conclude that the think you supposed is false. Since you supposed that the given statement was false, you now know that the given statement is true.

Proof by Contraposition

- Look to see if the statement to be proved is a universal conditional statement.
- If so, you can prove it by writing a direct proof of its contrapositive.

Find-the-Mistake Problems

All of the following problems contain a mistake. Identify and correct each one.

1. **Section 2.2**: The negation of "$1 < a < 5$" is "$1 \geq a \geq 5$."

2. **Section 2.2**: "P only if Q" means "if Q then P."

3. **Section 3.2**

    (a) The negation of "For all real numbers x, if $x > 2$ then $x^2 > 4$" is "For all real numbers x, if $x > 2$ then $x^2 \leq 4$."

    (b) The negation of "For all real numbers x, if $x > 2$ then $x^2 > 4$" is "There exist real numbers x such that if $x > 2$ then $x^2 \leq 4$."

    (c) The negation of "For all real numbers x, if $x > 2$ then $x^2 > 4$" is "There exists a real number x such that $x > 2$ and $x^2 < 4$."

4. **Section 3.2**: The contrapositive of "For all real numbers x, if $x > 2$ then $x^2 > 4$" is "For all real numbers x, if $x \leq 2$ then $x^2 \leq 4$."

5. **Section 3.3**: Statement: $\exists$ a real number x such that $\forall$ real numbers y, $x + y = 0$. Proposed negation: $\forall$ real numbers x, if y is a real number then $x + y \neq 0$.

6. **Section 4.1**: A person is asked to prove that the square of any odd integer is odd. Toward the end of a proof the person writes: "Therefore $n^2 = 2k + 1$, which is the definition of odd."

7. **Section 4.1**: *Prove:* The square of any even integer is even.

    Beginning of proof: Suppose that r is any integer. Then if m is any even integer, $m = 2r$. . . .

8. **Section 4.1**: *Prove directly from the definition of even:* For all even integers n, $(-1)^n = 1$.

    Beginning of proof: Suppose n is any even integer. Then $n = 2r$ for some integer r. By substitution, $(-1)^n = (-1)^{2r} = 1$ because $2r$ is even. . . .

9. **Section 4.1**: *Prove directly from the definition of even:* For all even integers n, $(-1)^n = 1$.

    Beginning of proof: Suppose n is any even integer. Then $n = 2r$ for some integer r. By substitution, $(-1)^{2r} = ((-1)^2)^r$. . . .

10. **Section 4.3**: *Prove:* For all integers a and b, if a and b are divisible by 3 then $a + b$ is divisible by 3.

    Beginning of proof: Suppose that for all integers a and b, if a and b are divisible by 3 then $a + b$ is divisible by 3. By definition of divisibility, ...

11. **Section 4.3**: *Prove:* For all integers a, if 3 divides a, then 3 divides a^2.

    Beginning of proof: Suppose a is any integer such that 3 divides a. Then $a = 3k$ for any integer k. . . .

12. **Section 4.3**: *Prove:* For all integers a, if $a = 3b + 1$ for some integer b, then $a^2 - 1$ is divisible by 3.

    Beginning of proof: Let a be any integer such that $a = 3b + 1$ for some integer b. We will prove that $a^2 - 1$ is divisible by 3. This means that $a^2 - 1 = 3q$ for some integer q. Then $(3b + 1)^2 - 1 = 3q$, and, since q is an integer, by definition of divisibility, $a^2 - 1$ is divisible by 3. . . .

13. **Section 4.4**: *Prove:* For all integers a, $a^2 - 2$ is not divisible by 3.

    Beginning of proof: Suppose a is any integer. By the quotient-remainder theorem with divisor $d = 3$, there exist unique integers q and r such that $a = 3q + r$, where $0 < r \leq 3$....

14. **Section 4.6**: *Prove by contradiction:* The product of any irrational number and any rational number is irrational.

    Beginning of proof: Suppose not. That is, suppose the product of any irrational number and any rational number is rational....

15. **Section 4.6**: The negation of "n is not divisible by any prime number greater than 1 and less than or equal to $\sqrt{n}$" is "n is divisible by any prime number greater than 1 and less than or equal to $\sqrt{n}$."

16. **Section 5.2**: The equation $1 + 2 + 3 + \cdots + n = \dfrac{n(n+1)}{2}$ is true for $n = 1$ because $1 + 2 + 3 + \cdots + 1 = \dfrac{1(1+1)}{2}$ is true.

17. **Section 5.2**: The equation $1 + 2 + 3 + \cdots + n = \dfrac{n(n+1)}{2}$ is true for $n = 1$ because

$$1 = \frac{1(1+1)}{2} \Rightarrow 1 = \frac{2}{2} \Rightarrow 1 = 1.$$

18. **Section 5.2**: *Prove by mathematical induction:* For all integers $n \geq 1$,

$$1 + 2 + 3 + \cdots + n = \frac{n(n+1)}{2}.$$

    Beginning of proof: Let the property $P(n)$ be

$$1 + 2 + 3 + \cdots + n = \frac{n(n+1)}{2} \text{ for all integers } n \geq 1....$$

19. **Section 6.1**: Given sets A and B, to show that A is a subset of B, we must show that there is an element x such that x is in A and x is in B.

20. **Section 6.1**: Given sets A and B, to show that A is a subset of B, we must show that for all x, x is in A and x is in B.

21. **Section 7.2**: To prove that $F: A \to B$ is one-to-one, assume that if $F(x_1) = F(x_2)$ then $x_1 = x_2$.

22. **Section 7.2**: To prove that $F: A \to B$ is one-to-one, we must show that for all x_1 and x_2 in A, $F(x_1) = F(x_2)$ and $x_1 = x_2$.

23. **Section 8.2**: Define a relation R on the set of all integers by $a\,R\,b$ if, and only if, $ab > 0$. To show that R is symmetric, assume that for all integers a and b, $a\,R\,b$. We will show that $b\,R\,a$.

Answers for the Find-the-Mistake Problems

All of the following problems contain a mistake. Identify and correct each one.

1. **Section 2.2**: The negation of "$1 < a < 5$" is "$1 \geq a \geq 5$."

   Answer: A statement of the form "$1 < a < 5$" is an *and* statement. Thus, by De Morgan's law, its negation is an *or* statement. The correct negation is $1 \geq a$ or $a \geq 5$.

2. **Section 2.2**: "P only if Q" means "if Q then P."

   Answer: "P only if Q" means that the only way P can occur is for Q to occur. This means that if Q does not occur, then P cannot occur, or, equivalently, "if P occurs then Q must have occurred," i.e., "if P then Q."

3. **Section 3.2**

   (a) The negation of "For all real numbers x, if $x > 2$ then $x^2 > 4$" is "For all real numbers x, if $x > 2$ then $x^2 \leq 4$."

   (b) The negation of "For all real numbers x, if $x > 2$ then $x^2 > 4$" is "There exist real numbers x such that if $x > 2$ then $x^2 \leq 4$."

   (c) The negation of "For all real numbers x, if $x > 2$ then $x^2 > 4$" is "There exists a real number x such that $x > 2$ and $x^2 < 4$."

   Answer to a, b, and c: The negation of a "For all" statement is a "There exists" statement, the negation of "if p then q" is "p and not q," and the negation of "$x^2 > 4$" is "$x^2 \leq 4$." The correct negation in all three cases is "There exists a real number x such that $x > 2$ and $x^2 \leq 4$."

4. **Section 3.2**: The contrapositive of "For all real numbers x, if $x > 2$ then $x^2 > 4$" is "For all real numbers x, if $x \leq 2$ then $x^2 \leq 4$."

   Answer: The contrapositive of "if p then q" is "if not q then not p." In this case p is $x > 2$ and q is $x^2 > 4$. Thus the correct answer is "For all real numbers x, if $x^2 \leq 4$ then $x \leq 2$."

5. **Section 3.3**: Statement: $\exists$ a real number x such that $\forall$ real numbers y, $x + y = 0$. Proposed negation: $\forall$ real numbers x, if y is a real number then $x + y \neq 0$.

   Answer: The proposed negation began correctly with "$\forall$ real numbers x," but the continuation should be the existential statement "$\exists$ a real number y such that $x + y \neq 0$."

6. **Section 4.1**: A person is asked to prove that the square of any odd integer is odd. Toward the end of a proof the person writes: "Therefore $n^2 = 2k + 1$, which is the definition of odd."

   Answer: For an integer to be odd means that it equals 2 times some integer plus 1. So it is not correct to say that "$2k + 1$ *is* the definition of odd." The person should have written: "Therefore $n^2 = 2k + 1$, where k is an integer, and so n^2 is odd by definition of odd."

7. **Section 4.1**: *Prove:* The square of any even integer is even.

   Beginning of proof: Suppose that r is any integer. Then if m is any even integer, $m = 2r$....

   Answer: To prove that the square of any even integer is even, you must start by supposing you have a *[particular but arbitrarily chosen]* even integer. By using the definition of even, you can *deduce* what the even integer must look like, namely that it must equal $2 \cdot$ (some integer). A correct proof would start with an even integer m and deduce the existence of an integer r such that $m = 2r$. This "proof" has it backwards.

8. **Section 4.1**: *Prove directly from the definition of even: For all even integers n, $(-1)^n = 1$.*

   Beginning of proof: Suppose n is any even integer. Then $n = 2r$ for some integer r. By substitution, $(-1)^n = (-1)^{2r} = 1$ because $2r$ is even....

   Answer: By claiming that $(-1)^{2r} = 1$, this "proof" assumes what is to be proved, namely that (-1) raised to an even power equals 1.

9. **Section 4.1**: *Prove directly from the definition of even: For all even integers n, $(-1)^n = 1$.*

   Beginning of proof: Suppose n is any even integer. Then $n = 2r$ for some integer r. By substitution, $(-1)^{2r} = ((-1)^2)^r$...

   Answer: The fact that $(-1)^{2r} = ((-1)^2)^r$ follows from a property of exponents; it is not true "by substitution." When you write "by substitution," you have to include the original variable in the equation that you write. Thus the following would be correct:
   Prove directly from the definition of even: For all even integers n, $(-1)^n = 1$.

   Beginning of proof: Suppose n is any even integer. Then $n = 2r$ for some integer r, and so

   $$\begin{aligned} (-1)^n &= (-1)^{2r} & \text{by substitution} \\ &= ((-1)^2)^r & \text{by a property of exponents...} \end{aligned}$$

10. **Section 4.3**: *Prove: For all integers a and b, if a and b are divisible by 3 then $a+b$ is divisible by 3.*

    Beginning of proof: Suppose that for all integers a and b, if a and b are divisible by 3 then $a + b$ is divisible by 3. By definition of divisibility, ...

    Answer: This proof begins by assuming exactly what is to be proved. If one assumes what is to be proved, there is nothing left to do!

11. **Section 4.3**: *Prove: For all integers a, if 3 divides a, then 3 divides a^2.*

    Beginning of proof: Suppose a is any integer such that 3 divides a. Then $a = 3k$ for any integer k....

    Answer: It is incorrect to say that "$a = 3k$ for any integer k" because k cannot be just "any" integer; in fact, the only integer that k can be is $k = a/3$. The correct thing to say is, "Then $a = 3k$ for some integer k."

12. **Section 4.3**: *Prove: For all integers a, if $a = 3b+1$ for some integer b, then $a^2 - 1$ is divisible by 3.*

    Beginning of proof: Let a be any integer such that $a = 3b + 1$ for some integer b. We will prove that $a^2 - 1$ is divisible by 3. This means that $a^2 - 1 = 3q$ for some integer q. Then $(3b + 1)^2 - 1 = 3q$, and, since q is an integer, by definition of divisibility, $a^2 - 1$ is divisible by 3....

    Answer: This "proof" assumes something equivalent to what is to be proved. After stating "We will prove that $a^2 - 1$ is divisible by 3" it is correct to state that ."This means that $a^2 - 1 = 3q$ for some integer q." However, the following sentence assumes that the integer q has been shown to exist, which is not the case.

13. **Section 4.4**: *Prove: For all integers a, $a^2 - 2$ is not divisible by 3.*

    Beginning of proof: Suppose a is any integer. By the quotient-remainder theorem with divisor $d = 3$, there exist unique integers q and r such that $a = 3q + r$, where $0 < r \leq 3$.

    Answer: The inequality is incorrect; it should be $0 \leq r < 3$.

14. **Section 4.6**: *Prove by contradiction:* The product of any irrational number and any rational number is irrational.

Beginning of proof: Suppose not. That is, suppose the product of any irrational number and any rational number is rational.

Answer: A proof by contradiction start with the negations of the statement to be proved. In this case, the statement to be proved is universal, and so its negation is existential. However, this proposed proof begins with a universal statement. A correct way to begin the proof is the following:

Beginning of proof: Suppose not. That is, suppose there exists an irrational number and a rational number whose product is rational.

15. **Section 4.6**: The negation of "n is not divisible by any prime number greater than 1 and less than or equal to $\sqrt{n}$" is "n is divisible by any prime number greater than 1 and less than or equal to $\sqrt{n}$."

    Answer: Consider negating the statement "He does not have any money." The negation is not "He does have any money," it is "He does have some money." Similarly, the negation of "n is not divisible by any prime number greater than 1 and less than or equal to $\sqrt{n}$" is not "n is divisible by any prime number greater than 1 and less than or equal to $\sqrt{n}$." It is "n is divisible by *some* prime number greater than 1 and less than or equal to $\sqrt{n}$," or "There exists a prime number greater than 1 and less than or equal to $\sqrt{n}$ that divides n."

16. **Section 5.2**: The equation $1 + 2 + 3 + \cdots + n = \dfrac{n(n + 1)}{2}$ is true for $n = 1$ because $1 + 2 + 3 + \cdots + 1 = \dfrac{1(1 + 1)}{2}$ is true.

    Answer: When $n = 1$, the expression $1 + 2 + 3 + \cdots + n = 1$; it does not equal $1 + 2 + 3 + \cdots + 1$.

17. **Section 5.2**: The equation $1 + 2 + 3 + \cdots + n = \dfrac{n(n + 1)}{2}$ is true for $n = 1$ because

$$1 = \frac{1(1 + 1)}{2} \Rightarrow 1 = \frac{2}{2} \Rightarrow 1 = 1.$$

Answer: A false statement can imply a true conclusion. So deducing a true conclusion from a statement is not a valid way to prove that the statement is true.

18. **Section 5.2**: *Prove by mathematical induction:* For all integers $n \geq 1$,

$$1 + 2 + 3 + \cdots + n = \frac{n(n + 1)}{2}.$$

Beginning of proof: Let the property $P(n)$ be

$$1 + 2 + 3 + \cdots + n = \frac{n(n + 1)}{2} \text{ for all integers } n \geq 1....$$

Answer: The job of a proof by mathematical induction is to prove that a given property is true for all integers greater than or equal to a given integer. In this example, the property $P(n)$ is simply the equation

$$1 + 2 + 3 + \cdots + n = \frac{n(n + 1)}{2},$$

and the proof by mathematical induction establishes that $P(n)$ is true for all integers $n \geq 1$. The mistake is including the words "for all integers $n \geq 1$" as part of $P(n)$ because these words make $P(n)$ identical with what is to be proved.

19. **Section 6.1**: Given sets A and B, to show that A is a subset of B, we must show that there is an element x such that x is in A and x is in B.

    Answer: This answer implies that for A to be a subset of B, it is enough for there to be a single element that is in both sets. But this is false. For instance, if $A = \{1, 2\}$ and $B = \{2, 3\}$, then 2 is in both A and B, but A is not a subset of B because 1 is in A and 1 is not in B. In fact, for A to be a subset of B means that for all x, *if* x is in A *then* x must be in B.

20. **Section 6.1**: Given sets A and B, to show that A is a subset of B, we must show that for all x, x is in A and x is in B.

    Answer: There are two problems with this answer. One is that it implies that A and B are identical sets, whereas for A to be a subset of B it is possible for B to contain elements that are not in A. In addition, because no domain is specified for x, it appears to say that everything in the universe is in both A and B, which is not the case for most sets A and B.

21. **Section 7.2**: To prove that $F\colon A \to B$ is one-to-one, assume that if $F(x_1) = F(x_2)$ then $x_1 = x_2$.

    Answer: Assuming that "if $F(x_1) = F(x_2)$ then $x_1 = x_2$" is essentially the same as assuming that F is one-to-one. In other words, it essentially assumes what needs to be proved.

22. **Section 7.2**: To prove that $F\colon A \to B$ is one-to-one, we must show that for all x_1 and x_2 in A, $F(x_1) = F(x_2)$ and $x_1 = x_2$.

    Answer: This statement implies that for all x_1 and x_2 in A, $x_1 = x_2$. In other words, it implies that there is only one element in A, which is very seldom the case.

23. **Section 8.2**: Define a relation R on the set of all integers by $a\,R\,b$ if, and only if, $ab > 0$. To show that R is symmetric, assume that for all integers a and b, $a\,R\,b$. We will show that $b\,R\,a$.

    Answer: The problem with these statements is that saying "assume that for all integers a and b, $a\,R\,b$" is equivalent to saying that every integer is related to every other integer by R. This is not the case. For instance, -1 is not related to 1 because $(-1) \cdot 1 = -1$ and $-1 \not> 0$.